PLANT PARENTING

plant
PARENTING

EASY WAYS TO MAKE MORE HOUSEPLANTS, VEGETABLES, AND FLOWERS

LESLIE F. HALLECK

TIMBER PRESS
PORTLAND, OREGON

Published in 2019 by Timber Press, Inc.
The Haseltine Building
133 S.W. Second Avenue, Suite 450
Portland, Oregon 97204-3527
timberpress.com

Printed in China

Text design by Anne Kenady
Cover design by Adrianna Sutton

ISBN 978-1-60469-872-5
Catalog records for this book are available from the
Library of Congress and the British Library.

CONTENTS

Cured succulent leaf cuttings rooting into mini terra cotta pots. Too cute!

Introduction

Have you caught the plant-keeping or gardening bug? If so, you might find yourself thinking a lot about how to make more of the plants you love. The spirit of nurturing ownership, thoughtful collecting, and creative display is in full bloom in the plant world. Houseplants, both old and new, are hot again with homeowners, among apartment and dorm dwellers, and at the office. Succulents, orchids, and unusual foliage plants fill the feeds of millions of social media users. Indoor and outdoor edible and ornamental gardening is gaining momentum, especially in tight, urban environments.

For amateur plant enthusiasts and professional horticulturists alike—myself included—caring for houseplants and starting seeds are often gateways to a full-fledged gardening addiction. This is a good thing. I still fondly remember getting my first houseplant as a gift when I was about seven years old: a bird's nest fern in a yellow pot. Keeping and caring for that one indoor plant sparked a lifelong interest in plants and a desire to make more of them any way I could. By college, I'd multiplied my indoor plant collection to the point where one could barely see out of the windows of the house I rented. It was my way of living with nature, indoors.

More than simply looking for ways to bring nature indoors or grow your own food, you might also want to collect and nurture plants, as well as connect with other plant keepers. As your plant addiction grows, so does your desire to cultivate, on your own, more of the plants you love instead of buying them already finished. Propagating your own plants is also a great way to stretch your plant buying and gardening budget.

If you're a new plant enthusiast, a young gardener, or an indoor and outdoor plant keeper who has never propagated your own plants before, this book will help you learn the basics of plant propagation. If you're a gardener who wants to grow your own edible and ornamental starts, expand your indoor plant collections, or even propagate your own citrus and fruit trees, I will teach you how. If you're a city dweller or balcony gardener who doesn't have yard space, or you need an introduction to propagation to stretch your plant-buying budget, you've also come to the right place.

This book will introduce you to the essential tools you need to start your plant propagation quest and demystify the art of basic seed and vegetative propagation techniques with easy-to-follow instructions. I've chosen to highlight certain plants that are easily accessible and make good examples for introductory propagation techniques. You can then take these techniques and apply them to many other plants of your choosing.

If you're new to plant collecting and propagation, you might be afraid you'll kill your seedlings or cuttings. Don't worry: *you will*. First-time failures can make you reluctant to try again. Fear of killing plants may even stop you before you get started, so don't let green guilt get the best of you. Killing a few plants, or even a lot of them, is ultimately how you learn to grow them successfully. Keep trying until you find the right recipe for success. Green thumbs are earned, not born.

Now that you've picked up this book, you're on your way to becoming a full-fledged plant parent! Welcome to the club.

Plants use a variety of reproductive strategies. African violet (*Saintpaulia* spp.) leaf-petiole cuttings, aeonium stem-tip cutting, and an avocado seed, each rooting in water.

How Plants

MULTIPLY

Understanding Basic Plant Reproduction, Propagation, and Names

Ready to multiply your plant collection? I'm sure you're eager to begin propagating a long list of your favorite or hard-to-find plants—or get those tomato seedlings started. First, let's do a quick overview of how different plants multiply. The method you use at home to make more of the plants you love will depend on the type of plant you want to propagate and how it grows and flowers. Not all plants can be propagated the same way or under the same growing conditions. If you've struggled to get your succulents to root before they rot or can't seem to get your lettuce seed to germinate, we're going to get you on the right track.

Depending on the type of plants and their natural environment, plants typically fall into two propagation groups: seeds and vegetative cuttings (also called clones). Some plants are more easily propagated by seed and others by cuttings. Some plants are easily propagated using both methods. Most first-time plant propagators begin with seed starting and water rooting, and then move on to more varied types of plant cloning techniques.

Many plants have seeds that are very easy to germinate, while others can be downright stubborn. Some plants will root almost overnight in just water, while others may quickly rot. Non-flowering plants, such as ferns, do not make seeds at all; rather, they multiply by spreading spores, and can also be propagated

∧ The cells of this African violet leaf petiole can grow new root and shoot tissue, creating an entirely new plant. Not all plants have this potential.

vegetatively. The trick is to start with the right plants and the right methods, so you can quickly learn the ropes and have early success. Then, once you feel more confident, you can graduate to propagating more plants.

While we won't get into any heavy botany here, it's important to know that different plant parts—such as a seed, stem, or leaf—either have, or lack, the ability to grow new types of plant tissue, such as new roots or new shoots. That's why you can use certain parts of a plant for cuttings, but not others.

< Starting seeds is an inexpensive way to make a lot more of the plants you love.

These bees are busy feeding on nectar from pincushion flowers (*Scabiosa* spp.). They are also moving pollen between the flowers.

STARTING WITH SEEDS

～～～

Propagating plants from seed is relatively simple and saving seed from your harvest can be an easy and inexpensive way to grow more of your favorite flowers and food. In flowering plants, seed production involves cross-pollination of a flower, the fertilization of the female ovary with male pollen, and the subsequent development of a fruit and seed, which grows a new plant once it germinates.

In the outdoor garden, wind and pollinators such as bees, wasps, butterflies, and a multitude of other insects aid the pollination process. Insects are attracted to flowers as a source of food. While visiting, pollinators pick up pollen on their bodies. The insects then move the pollen around on the flower or transport it to other nearby flowers. It's quite fun to watch and a great alternative to TV.

∧ Mature butterfly weed (*Asclepias tuberosa*) seeds are dry and ready to save or sow.

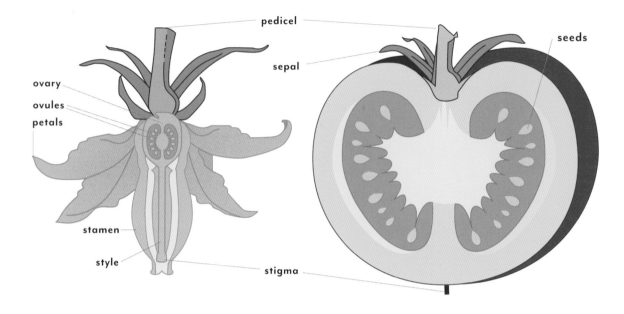

FLOWER   FRUIT

pedicel

seeds

sepal

ovary

ovules

petals

stamen

style

stigma

︿ A perfect (or self-pollinating) flower has all the parts it needs to develop fruit and set seeds.

Some flowers have both female parts and male parts. These flowers are called perfect or self-pollinating. A little wind or a light shake is all that is required for pollen to move around and pollinate the flower—but pollinators certainly help. Plants with perfect flowers typically set a lot of seed and fruit easily, so they are great for beginner gardeners and seed collectors. Think cosmos flowers and pepper plants.

Some plants have separate male and female flowers on the same plant. These flowers are called incomplete. Squash and cucumber are examples of plants with incomplete flowers. The pollen from the male flower must travel to a nearby female flower on the same plant for pollination to occur. These crops need a helping hand from pollinators (or you) to move pollen from male flowers to female flowers so that fruit and seed can develop.

Seed from some types of plants can be difficult to collect or germinate. The seeds of certain succulents, for example, are so tiny that saving and germinating them can be challenging, regardless of whether you're new to plant propagation or an experienced grower. For plants with tiny seeds, you'll need to use care and patience to germinate them successfully and grow them into mature plants. If you don't have the patience to wait for seedlings, you can propagate many succulents and cacti by removing small offsets (pups) that will grow around the base of the mother plant. (We call it a mother plant simply because it provides the material to grow the new baby plants.)

< Love-in-a-mist (*Nigella damascena*) has perfect flowers that contain both female and male parts.

< The seed pods of love-in-a-mist are quite beautiful and easy to collect; and the bounty of dark black seeds spreads easily around the garden without your help.

∧ Some species have plants with only male flowers or only female flowers. Only female plants of holly plants produce berries. Once their female flowers are pollinated by pollen from a nearby male plant, berries can develop.

Squash plants produce separate male and female flowers. Baby fruit develops from the ovary, situated just behind a female flower. So, when you see a tiny baby fruit, you know it's a female flower. >

When should you choose seed propagation?

- When the plant you want to grow reproduces easily from seed

- When you don't have access to vegetative cuttings or a mother plant

- When you want to produce the most plants for the lowest cost

- When you have time to wait for seedlings to mature

- When growing a cutting would deform the mother plant's growth habit

- When you are growing microgreens or sprouts

∧ The seeds of *Mammillaria* cactus and living stone succulents (*Lithops* spp.) are no bigger than a speck of dust, are moderately challenging to germinate, and require many years to achieve a large size. Here, tiny *Mammillaria* cactus (at two stages of early germination) and living stone seedlings several weeks after germination.

VEGETATIVE PROPAGATION

~

When you are unable to collect seed from your chosen plant or the seeds are innately difficult to collect, purchase, or sow, vegetative propagation, also called cloning, is a better option. Cloning involves using existing plant tissue, such as a stem, leaf, petiole (the tissue that connects the leaf to the plant stem), root tissue, or young plantlet from a mother plant. Many plants can generate new root tissue (adventitious roots) and new shoot tissue (adventitious shoots) from other parts of the plant such as a stem or leaf. Once the new roots and shoots form from the existing plant tissue, a new plant clone develops. The resulting new plant will be identical to the mother plant from which you took the cutting.

New adventitious roots and buds can develop at the base of a leaf, leaf petiole, stem internode, node, or at the base where a stem has been cut, depending on the plant species. Have you ever been instructed to plant tomatoes deeper in the soil so that you bury part of the stem? That's because tomatoes grow adventitious roots along their stems, which can make the plants stronger and more vigorous.

A callus often develops at the base of a cutting at the site of a wound, when conditions are good for rooting. A callus will look like a knot or knuckle growing at the base of a cut stem or leaf. Some plants will develop a callus and adventitious roots, or crown roots, simultaneously; other plants, such as citrus and peppers, must develop a callus first, before new roots can grow.

Many succulents can generate new root tissue from the base of a fallen leaf. The leaf will typically form a callus, and then new adventitious roots will form, followed by an adventitious bud and new shoot.

Other plants, such as the popular houseplants peperomia and begonia, can develop new roots on the leaf petiole and even from the veins along the leaf itself. These types of plants are masters of multiplication and can be propagated in several different ways. However, don't expect your tomatoes or citrus plants to sprout new roots from just a leaf; the cells in their leaf and petiole tissue can't grow new root or shoot tissue.

Some plants develop offsets, or pups, which come complete with their own root tissue and then can be

< I accidentally broke this piece of African violet crown away from the rest of the plant. I supported it in water and re-rooted the crown, making a clone of the mother plant.

< You can clearly see the adventitious roots growing along the stems of this potted tomato plant. These stems are good candidates to provide cuttings, since root tissue is already growing.

∧ This pepper cutting has developed a knuckle-like callus at the base of the cut stem; crown roots (wound adventitious roots) will soon develop.

∧ These crown roots developed on a citrus cutting after the stem was cut and then held in an aeroponic propagator for several weeks.

divided from their mother and immediately potted up. Airplane plant (also called spider plant) produces easy-to-propagate offsets on flowering stems, while other plants grow offsets from under the soil around the base of the mother plant, or even on the edges of the plant's leaves. Strawberry plants develop offsets on runner stems (stolons). These plantlets can be snipped off the stem and planted to grow entirely new plants.

In addition to green plant parts that can be used for vegetative propagation, other plant tissues allow plants to multiply vegetatively, such as bulbils, bulbs, bulblets, corms, crowns, rhizomes, stolons, suckers, and tubers. Many flowering plants can multiply by seed, as well as vegetatively using these other parts.

Ultimately, you'll need to get to know the plant you're propagating, learn how it reproduces in nature, and find out which plant parts are best used for propagation. Feel free to try several methods for a given plant to see which one works best for you.

∧ An adventitious bud and shoot developed at the base of this echeveria leaf after I placed it in a container with potting mix.

A fallen echeveria leaf develops new adventitious roots at its base, followed by an adventitious bud and shoot, which then develop into a clone of the original mother plant.

New adventitious roots and a bud and shoot have developed at the base of this peperomia leaf petiole.

Airplane plants (*Chlorophytum comosum*) are easy to propagate from the offsets they produce at the ends of flower stems. >

Chinese money plant (*Pilea peperomioides*) is a popular houseplant that can be propagated by removing offsets, or pups, that emerge from the soil on the main stem. >

Tissue Culture

Tissue culture is a type of propagation performed in test tubes or Petri dishes. These days it's more commonly referred to as micro-propagation. It's a fascinating process by which tissue is extracted from a plant and grown in a sterile contained environment in a nutrient solution. The tiny propagule (plantlet) that forms is referred to as an explant. As the explant grows, it is transplanted into more traditional growing media and containers. Tissue culture is typically used in the commercial industry to propagate plants that are difficult to reproduce by other methods, to reduce disease and pest issues, and to produce large quantities of plantlets. However, some tissue culture kits are now available online to the home gardener, as well as books and online information that go into detail on tissue culture methods. As always, I encourage you to experiment beyond the boundaries of this book if you are so inspired.

∧ Tissue-cultured plants growing in nutrient solution can be cloned in large quantities with fewer pest and disease problems.

When should you choose vegetative propagation?

- · When the seed of a plant is difficult to collect or not available
- · When the seed of choice is very expensive
- · When the plant you want to grow does not produce seed
- · When growing a mature plant from seed will take longer than you want to wait
- · When you want to reproduce clones from a hybrid cultivar
- · When you want to create an exact copy of a plant variety you already have

Common Propagation Methods for Some Popular Plants

Key: **SS** = seeds or spores, **SC** = stem-tip cutting, **CC** = cane cutting, **LC** = leaf cutting, **OR** = offsets, runners, **SK** = suckers, **R** = rhizomes, bulbs, bulblets, tubers, corms, or root cuttings, **B** = bulbils, **PB** = pseudobulbs, **L** = layering, **AL** = air layering, **D** = division, **TC** = tissue culture

PLANT	SPECIES	SS	SC	CC	LC	OR	SK	R	B	PB	L	AL	D	TC
African violet	*Saintpaulia* spp.		x		x	x							x	x
agave	*Agave* spp.	x				x			x				x	
ajuga	*Ajuga reptans*		x			x								
aloe	*Aloe* spp.					x							x	
apple	*Malus* spp.	x	x								x	x		
banana	*Musa acuminata*					x	x						x	
basil	*Ocimum basilicum*	x	x										x	
bean	*Phaseolus vulgaris*	x												
beard tongue	*Penstemon* spp.	x	x										x	
begonia	*Begonia* spp.	x	x		x			x					x	x
blackberry	*Rubus* spp.		x			x	x				x			
black-eyed Susan	*Rudbeckia hirta*	x											x	
blanketflower	*Gaillardia* spp.	x												
bleeding heart	*Lamprocapnos spectabilis*												x	
bluebonnet	*Lupinus* spp.	x												
boat orchid	*Cymbidium* spp.	x								x			x	x
Boston fern	*Nephrolepis exaltata*	x											x	x
broccoli	*Brassica oleracea*	x												
butterfly bush	*Buddleja* spp.		x								x			
candytuft	*Iberis* spp.	x	x										x	
canna lily	*Canna indica*	x						x					x	
cardinal flower	*Lobelia cardinalis*	x	x										x	
cast iron plant	*Aspidistra elatior*							x					x	

PLANT	SPECIES	SS	SC	CC	LC	OR	SK	R	B	PB	L	AL	D	TC
cattleya orchid	*Cattleya* spp.					x		x		x			x	x
cauliflower	*Brassica oleracea*	x												
Chinese evergreen	*Aglaonema commutatum*		x	x									x	
Chinese money plant	*Pilea peperomioides*		x			x							x	x
citrus	*Citrus* spp.	x	x									x		
clematis	*Clematis* spp.		x								x			
coleus	*Solenostemon* spp.	x	x										x	x
columbine	*Aquilegia* spp.	x											x	
coneflower	*Echinacea* spp.	x						x					x	
coral bells	*Heuchera* spp.												x	x
corn plant	*Dracaena* spp.		x	x								x		
cosmos	*Cosmos* spp.	x												
creeping phlox	*Phlox subulata*		x			x							x	
crocus	*Crocus sativus*	x						x					x	
croton	*Codiaeum variegatum var. pictum*		x										x	
daffodil	*Narcissus* spp.	x						x					x	
dahlia	*Dahlia* spp.		x					x					x	
dancing lady orchid	*Oncidium* spp.					x				x			x	x
daylily	*Hemerocallis* spp.	x						x					x	x
dendrobium	*Dendrobium* spp.	x				x				x			x	x
dieffenbachia	*Dieffenbachia* spp.		x			x							x	
echeveria	*Echeveria* spp.	x	x		x								x	
eggplant	*Solanum melongena*	x	x											
evening primrose	*Oenothera speciosa*	x											x	
fern	many species	x								x			x	x
ficus	*Ficus benjamina*		x									x		
fiddle leaf fig	*Ficus lyrata*		x									x		
four o-clocks	*Mirabilis jalapa*	x												

PLANT	SPECIES	SS	SC	CC	LC	OR	SK	R	B	PB	L	AL	D	TC
garlic	*Allium sativum*	x						x	x				x	
gaura	*Gaura* spp.	x											x	
geranium	*Pelargonium* spp.	x	x											
gerbera daisy	*Gerbera* spp.	x												
ginger	*Zingiber officinale*							x						
gladiolus	*Gladiolus* spp.							x						
hens and chicks	*Sempervivum* spp.	x			x	x							x	
hibiscus	*Hibiscus rosa-sinensis*		x											
hosta	*Hosta* spp.	x						x					x	x
iris	*Iris* spp.	x						x					x	
Jerusalem sage	*Phlomis fruticosa*		x											
lamb's ear	*Stachys byzantina*												x	
lavender	*Lavandula* spp.	x	x										x	
Lenten rose	*Helleborus* spp.												x	
lettuce	*Lactuca sativa*	x												
lilac	*Syringa vulgaris*		x								x		x	
lily	*Lilium* spp.	x				x		x	x				x	
living stones	*Lithops* spp.	x												
Mexican hat	*Ratibida columnifera*	x												
milkweed	*Asclepias* spp.	x	x											
mondo grass	*Ophiopogon japonicus*	x											x	
moneywort	*Lysimachia nummularia*		x			x							x	
moth orchid	*Phalaenopsis* spp.			x									x	x
mother of thousands	*Bryophyllum diagremontianum*		x	x		x			x				x	
mum	*Chrysanthemum* spp.		x			x							x	
night-blooming cereus	*Cereus* spp.	x	x										x	
onion	*Allium* spp.	x							x					
ornamental grasses	many species	x											x	

PLANT	SPECIES	SS	SC	CC	LC	OR	SK	R	B	PB	L	AL	D	TC
peace lily	*. spp.*												x	
peperomia	*. eperomia* spp.		x		x	x								
peppermint	*Mentha* spp.		x			x							x	
pepper	*Capsicum* spp.	x	x											
philodendron	*Philodendron* spp.		x			x							x	
pincushion flower	*Scabiosa* spp.	x											x	
plumbago	*Ceratostigma plumbaginoides*					x								
potatoes	*Solanum tuberosum*							x						
pothos ivy	*Pothos* spp.		x			x							x	
rose	*Rosa* spp.		x								x			x
rosemary	*Rosmarinus officinalis*	x	x										x	
sage	*Salvia officinalis*		x										x	
shasta daisy	*Leucanthemum* spp.							x					x	
snake plant	*Sansevieria* spp.		x		x			x					x	
speedwell	*Veronica* spp.												x	
St. John's wort	*Hypericum perforatum*		x											
stone crop	*Sedum* spp.	x	x										x	
tomato	*Solanum lycopersicum*	x	x											
tulips	*Tulipa* spp.	x						x					x	
vanda orchid	*Vanda* spp.		x	x		x							x	
wax flower	*Hoya* spp.		x										x	
wormwood	*Artemisia* spp.		x									x	x	
zebra plant	*Haworthia* spp.					x							x	
zinnia	*Zinnia* spp.	x												

UNDERSTANDING PLANT NAMES

Along with understanding how plants multiply, it's useful to understand how plants are named and categorized so you use the right propagation method. While you don't need to be an expert on this subject to make more plants, understanding a few botanical terms related to propagation will prove helpful—especially if you plan to collect and save your own seed. Varieties, open-pollinated cultivars, hybrids, heirlooms, and GMO/GE plants are all types you may grow at some point and terms you'll see printed on seed packets or plant tags. These terms are often used incorrectly and interchangeably, which can be a source of major confusion for home gardeners. So, let's break them down.

binomial species name that includes a genus and a specific epithet. Common names are the names we make up for plants, so we can refer to them without knowing the species name. For example, *Pilea peperomioides* is the species name for a popular houseplant native to southern China, which is known by the common name Chinese money plant. Sometimes the common name for a plant is simply the genus. Know that common names vary widely depending on where you live—so most plants can have several different common names. *Sansevieria* spp., for example, are often called snake plant, mother-in-law's tongue, devil's tongue, and snake's tongue.

SPECIES AND COMMON NAMES

A plant species is a group of plants in which two individuals can produce fertile offspring. Plants are given a

VARIETIES

A variety of a plant species is one that may show slight differences in physical characteristics from the original plant species and can occur naturally in the wild—with or without human aid. Varieties can result from

Dendrobium ×hybrida 'Precious Pearl' orchid is a hybridized cultivar that produces a profusion of stunning flowers.

natural mutations because of differing environmental conditions, or from a seedling that arises from naturally occurring cross-pollination within the same species. Varieties of a species may have variegated foliage, a slightly different flower color, size, or other such physical difference, but it is not significant enough to classify the variety as a new species. People can sometimes spot and select a naturally occurring variety and sell it commercially with a given variety name.

You'll see varieties noted with the designation "var." within a species name. For example: the species name of heartleaf philodendron, a popular vining houseplant, is *Philodendron hederaceum*. A naturally occurring variety of this plant is *P. hederaceum* var. *oxycardium*, which has glossier leaves than the original species.

⌃ Heartleaf philodendron (*Philodendron hederaceum*), is a popular species of houseplant.

HYBRIDS AND CULTIVARS

Hybridizing involves an exchange of pollen from two different varieties of a plant species, or closely related species within the same genus, each having desirable characteristics. The hybrid that grows from the seed that resulted from that pairing will exhibit a mix of characteristics from both parent plants. While hybridization occurs all the time in nature—by wind, animals, and pollinators carrying pollen between plants—the type of crosses made by people generally have a much lower probability of occurring in the wild. When humans intentionally make such a hybrid by manually cross-pollinating two plants, we call the result a hybrid cultivar. You'll see hybrids designated by a cultivar name, noted in single quote marks, such as *Gomphrena* ×*hybrida* 'Pink Zazzle'.

The first generation from such a human-made plant cross is called the F1 generation. F1 hybrids are typically the most vigorous and display the best characteristics, such as bigger fruit or better disease resistance. However, if you collect seeds from an F1 hybrid they won't come true to type; meaning they can express a variety of different characteristics from the parent plants, so you never quite know what you're going to get.

To recreate an F1 hybrid, you must re-cross the original two parent varieties to create the same F1 hybrid, propagate the hybrid cultivar vegetatively, or grow it from tissue culture in a lab. Therefore, F1 hybrid cultivars typically cost more to purchase than species or varieties.

Sansevieria Moonshine

Sansevieria Black Gold Compacta

Sansevieria Bantell

∧ There are about seventy different species of snake plant (*Sansevieria* spp.) and countless varieties and cultivars.

< A chartreuse-leaved variety of heartleaf philodendron named *Philodendron hederaceum* var. *aureum*. You'll sometimes see it called 'Lemon Lime' or 'Neon'.

∧ Species of gomphrena (globe amaranth) typically produce smaller flowers than this hybridized gomphrena with the cultivar name 'Pink Zazzle'.

This tiny viola seedling that seeded itself into the cracks in my driveway was produced from a dark blue viola hybrid cultivar that I planted in my garden. This baby viola plant has a different growth habit and a lighter flower color than its F1 hybrid parent. ＞

∧ When you buy this cultivar of philodendron, you should see the patented cultivar name 'Brasil' on the tag.

∧ The only way to reliably reproduce the same cultivar of cannabis is to take vegetative cuttings. If you collect and start seed from these plants, you'll end up with a wide variety of characteristics in your seedlings.

Some cultivars originate as result of a natural mutation, also called a "sport," of a plant variety. These cultivars made it into the marketplace because they were selected by a person and reproduced under a specifically marketed cultivar name. These plants are called selected cultivars. Remember the heart-leaf philodendron I mentioned previously? Someone growing *Philodendron hederaceum* var. *oxycardium* lucked into a sport of the variety that had interesting chartreuse variegation on the foliage. That person then selected the variegated beauty, propagated it vegetatively, and patented it under the cultivar name 'Brasil'. So, the name on the plant tag for the hybrid would read *Philodendron hederaceum* var. *oxycardium* 'Brasil'.

BRANDED PLANT NAMES

Be aware there are also trade names for plant cultivars for marketing purposes. If a company that is going to distribute and sell a certain plant cultivar doesn't like the assigned cultivar name, it may create a new brand name that better fits the marketing strategy. Ever heard of the popular Peace rose? You may have seen tags that list the plant as *Rosa* 'Peace'—but that would be incorrect. 'Peace' is not the real cultivar name—it's just a trade brand name used for marketing purposes. The real cultivar name for hybrid Peace rose is 'Madame A. Meilland'. While trade names are helpful for increasing plant brand exposure, they can make things more difficult for consumers and enthusiasts when it comes to searching for the right plant.

OPEN-POLLINATED PLANTS

Seed produced from naturally occurring flowering plant species or varieties is true to type, meaning that you'll get roughly the same characteristics from seeds collected and grown from the original variety as you will with varieties that are typically open-pollinated. An open-pollinated variety can self-pollinate or be pollinated by another specimen of the same variety and produce essentially the same variety from seed. Depending on the plant, you can also vegetatively reproduce (clone) the variety to make sure you keep the same characteristics and benefits expressed by the variety.

If you want to collect and store your own seed from plants that you are growing—and grow approximately the same plant from which you saved the seed—you can only do so with open-pollinated plants. If you choose to save and sow seed from hybrid plants, you'll end up growing seedlings that show a variety of different characteristics, which is a fun way to experiment and to breed new plants. If a given selected cultivar also happens to be open-pollinated, then it's possible for you to collect true-to-type seed from such a cultivar.

One word of caution: open-pollinated doesn't mean isolated pollination. Open-pollinated plants can still cross-pollinate with other closely related varieties if they are grown too close together, especially if they have incomplete flowers, or male and female flowers on separate plants (which means pollen must travel). Plant two different open-pollinated varieties of cucumber in the same backyard and they can cross-pollinate. The fruit they bear, plus the seeds you save, won't necessarily be uniform. If pollen can move from plant to plant via insects, wind, or air moved by fans, then there is potential for cross-pollination. If you can, group the same variety of open-pollinated plants close together or stick to self-pollinating plants (with perfect flowers) if you want to save seed.

HEIRLOOMS

To be designated an heirloom, a plant must be open-pollinated and have been in human cultivation for at least fifty years. Because heirlooms produce seed that grows true to type, these plants have been saved and passed down from generation to generation. You can reliably collect and save seed from your heirloom plant types, and be confident you'll grow the very same plant, with the same characteristics, from those seeds the next season.

Some gardeners prefer a plant variety be in cultivation for closer to 100 years to be considered a true heirloom. Therefore, younger heirlooms may be referred to as modern heirlooms. 'Green Zebra' tomato is a good example of a modern heirloom (and a selected cultivar).

GENETICALLY MODIFIED PLANTS

While often used interchangeably (erroneously in my view), the word *hybrid* is not synonymous with the term GMO (genetically modified organism) or the more accurate term GE (genetically engineered). Genetic engineering of plants involves altering a gene or multiple genes to create an abnormal trait. Genetic engineering can also involve introducing genes from a completely different organism, such as a bacterium, into the plant. Altering genes changes the biochemistry of the plant with the intent of getting it to do, produce, or resist something it naturally would not. So, when a plant or seed is labeled as a GMO or GE plant, that means its genes have been altered. A scientist *could* take a hybrid plant and genetically engineer it . . . but that does not mean all hybrids are GE or GMO. When people mistakenly equate hybrid seeds and plants with GMO/GE seeds and plants, it can cause unnecessary fear about growing many hybrid plants.

∧ When you grow heirloom varieties of artichoke, you can let some of the flower heads bloom and go to seed instead of harvesting them to eat. Save the seed to start the next season. Plus, the flowers are beautiful!

While the technical rules of botanical nomenclature are much more nuanced and complicated than I've described here, these basic definitions will help you make better choices about the plants you purchase and how you collect and save seed for your future edible crops and plant collections.

Legalities of Propagation

It might surprise you to learn that many plant cultivars are patented. Just like with photography, writing, art, or inventions, the concept of intellectual property also exists within the plant breeding world. It is illegal for you as a home gardener, or anyone in the trade, to propagate (either by seed or vegetative cuttings) and give away or sell plants that are protected by an active patent or have a patent pending—unless you are a grower with a license to grow and sell the variety or cultivar. The plant label you purchase with your original plant will display either a current patent number or the acronym PPAF, which stands for "Plant Patent Applied For."

As a home gardener, you *are* legally allowed to propagate patented plants purely for your own personal use. If the plant variety or cultivar you're collecting seed from or taking cuttings from has no active patent, then it is fair game for anyone to propagate, sell, or give away.

Propagating and selling patented plants for your yard sale or school fundraiser is a no-no. Make sure you are only propagating and gifting non-patented plants. Selling plants out of your home or online also typically requires special nursery licenses and property and plant inspections. Be sure to do your research first before you start selling or giving away plants you've propagated in your home and garden. Without all the hard work plant breeders put into their craft, we wouldn't have so many amazing plant varieties!

'Green Zebra' tomatoes are a reliable open-pollinated selected cultivar, and modern heirloom, with sweet green fruit. I can save seeds from season to season. >

∧ Several professional plant companies sell the hybrid rhizomatous begonia cultivar 'Phoe's Cleo', but it is not currently patented. That means you are allowed to both propagate and distribute it freely.

MATERIALS
and
TOOLS

**Everything You Need
to Propagate and Light
Your Plants**

When it comes to the materials and tools you'll need to start seeds and take plant cuttings, there are a wide variety of suitable options. You can often recycle containers from your kitchen or use simple household tools you already have on hand. There are also many inexpensive simple gardening tools that can make propagation easier. However, high-quality hand tools, plant grow lights, and automated propagators are often worth the investment if you plan to do a lot of propagating and gardening. The following materials and tool recommendations will help you get started with simple and ambitious propagation projects alike.

∧ A basic set of small hand tools for making more of the plants you love.

HAND TOOLS

You'll find that when it comes to propagation hand tools, bigger isn't always better. Mini-gardening tools are your friends. Precision is key, so look for hand tools that are small, sharp, and can squeeze into tight spaces to place tiny seeds, snip excess seedlings, or transplant them gingerly. Here are the most helpful basic hand tools that you'll use most often.

MINI-TRANSPLANTING TOOLS I love inexpensive mini-dibbers, trowels, and garden forks, which are perfect for making planting holes just the right size for new seedlings (dibber), as well as gently lifting and planting them (trowel). The roots of small seedlings are delicate and it's easy to break them if you try to lift them from the soil with your fingers. You can find wood and metal versions of all these tools.

SOIL SCOOP A soil scoop makes fast work of filling pots and trays. I like to keep a bin of potting soil and small soil scoops at the ready for filling seed trays or pots.

SEEDER Using a hand seeder can help you better place seeds and reduce thinning, so you don't waste seeds. They are also helpful when handling very small seeds that can be hard to see. They have tiny nozzles of different sizes and a squeezable compartment for seed loading.

TWEEZERS You can use tweezers to place seeds exactly where you want them or to lift out small plantlets and seedlings.

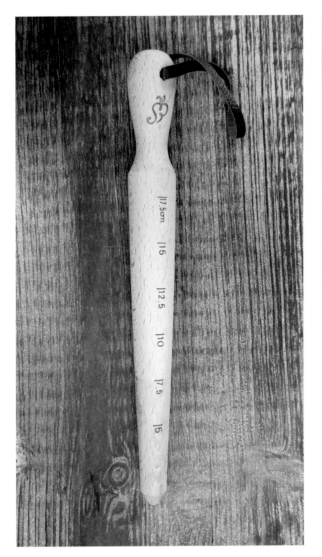

SPRAY AND SQUIRT BOTTLE You'll need a spray bottle that delivers water from a fine mist to small water droplets to mist seeds, delicate seedlings, and rooting cuttings. When you don't want to get water on the foliage of your young transplants (especially succulents), a squirt bottle that delivers a narrow stream of water directly to the root zone is handy.

WATERING CAN Keep a small watering can handy to water small seedlings, cutting plugs, and transplants.

RAZOR BLADE OR KNIFE When you are taking small root cuttings or divisions from plants, you may need a sharp razor blade or knife to remove them with precision from the mother plant.

DIBBER Dibbers are simple planting tools made of wood, plastic, or metal that are narrower at the bottom than at the top. Many come with depth measurement markers. They are handy for making a perfectly sized dent in your growing media when transplanting seedlings and cuttings.

PRUNERS AND SNIPS If you're taking cuttings from woody or perennial plants in the garden, you'll need a sturdy pair of bypass hand pruners that can cut stems up to 1 inch in diameter. For culling extra seedlings because too many have germinated, you'll need sharp snips (shears) with small blades. You can also use snips for taking soft-tissue cuttings from mother plants.

SOIL BLOCKER Soil blockers create soil plugs without the use of a container. Mix your desired seed mix, then make a square form with the soil blocker. Soil blockers can help you cut down on plastic or material waste.

PAPER POT MAKER These wooden tools form newspaper into functional pots for your seedlings and cuttings.

CLEANING BRUSH OR DUST BLOWER Keep small (½- to 1-inch) brushes or a small dust blower on hand to gently remove soil from plant leaves and seedlings.

CONTAINERS

~

Containers specifically made for starting seeds and cuttings range in size, shape, color, and material. Some are better for seeds; some are better for cuttings. Prefabricated seed-starting and watertight cutting trays are already segmented into small cells or plugs for you. You'll most commonly find seed-starting kits and plug trays sold in the twenty-five to fifty-cell count size.

I regularly use plug trays for starting a mixture of different plant varieties. Just because you have a fifty-cell plug tray, doesn't mean you have to start fifty tomato plants. You can start ten seeds each of five different types of vegetables or flowers in the same tray, or ten seeds each of five different tomato varieties if you really love tomatoes! Do make sure the seeds you start together do well in similar growing conditions. You can start seeds of warm-season crops such as tomatoes, peppers, and eggplant in the same tray or groups of cool-season veggies such as broccoli, cauliflower, and kale together. Remember to label your rows or the individual cell, so you won't forget what you planted where.

You can also find some smaller options that hold six to twelve cells or plugs. I typically use the smaller trays to start specialty plants such as succulents or a handful of cuttings. You don't have to use prefabricated seed trays—individual small plastic pots, biodegradable plug packs, small clay pots, or any number of other options will also work well.

Container size is very important when starting new seeds and cuttings. When choosing a container for starting your seeds, look for one that is no more than 2 to 3 inches deep. Larger containers that hold more soil may cause problems for tiny seedlings, as they can hold too much moisture or wick moisture away from the still-small root system.

The container in which you directly grow the seedling or cutting will typically have a drainage hole or should be porous. That means water will drain out of the drainage hole. Seedlings and cuttings grown in porous containers or plugs can be placed in solid trays to catch water runoff and hold a bit of extra moisture and humidity as the roots mature.

If you are careful not to overwater you can also grow seeds and cuttings in containers without drainage holes. I've repurposed many such containers for a variety of propagation needs. For example, I've found that silicone ice cube and baking molds make handy seed and cutting trays. The plugs pop right out of the

︿ Small 2-inch clay pots with a drainage hole are perfect for starting individual succulent and cactus seeds or for succulent cuttings.

︿ A small twelve-cell plug tray, with insert, watertight tray, and humidity dome.

︿ Porous seedling plugs placed in a watertight tray.

containers after they are rooted. I use the larger ones for cuttings and the smaller ones for tiny cactus and succulent seeds.

You can grow microgreens in a watertight seedling tray or a variety of other shallow containers without drainage holes, as you'll be harvesting the seedlings in short order.

Before you choose the type of container in which to start your seeds or cuttings, decide if you want to reuse the container. A container made from clay or plastic, for example, will be more stable and can be reused many times. Biodegradable containers and seed plugs, on the other hand, allow you to plant the entire cell directly into a larger pot or your garden without lifting out the seedling or disturbing its root system. The materials break down naturally in the soil.

Compressed fiber seedling pots can be planted directly into the ground or a larger container of soil. Simply fill them with soil and drop in seeds or stick cuttings. You don't need to remove the seedling or cutting once it's ready to be transplanted, as the containers are biodegradable and plant roots will work their way through the fiber. Just make sure to tear away the top of the container that is sticking out above the soil line.

When you want to be as thrifty or sustainable as possible in your propagation practices, you can use recycled materials or make your own pots and plugs. There are many recycled materials you may have on hand—egg cartons, egg shells, plastic containers from the grocery store, clear plastic cups, Styrofoam cups, and the like—that work well for starting seeds and cuttings. Shallow bowls and planters can be used to germinate seeds and grow microgreens. You can also create DIY containers and seed plugs.

∧ Silicone molds can pull double duty as propagation trays.

∧ Classic plastic nursery pots come in a range of sizes and colors. These small 2-inch and 4-inch diameter pots are perfect for starting seeds and cuttings. They can be washed and reused or recycled when they've run their course.

These compressed fiber seeding trays and pots can be filled with growing media, then separated and planted directly into the ground or larger pots so you don't have to disturb plant roots. >

< Sowing pea seeds into eggshells filled with potting mix. After seeds germinate, you can transplant the entire shell into a new container or into the garden.

∧ Tiny clay pots used for germinating succulent and cactus seeds are placed inside a repurposed plastic food container, with a piece of capillary mat to regulate moisture.

Available in many sizes, fabric pots are good for growing edible plants because they allow plant roots to breathe and can be washed and reused. ⟩

You can use a paper pot maker to create your own small, biodegradable pots. Simply cut a 3- by 10-inch section of newspaper and wrap it around the pot maker handle, and tuck under the bottom edges. Then press the paper into the pot maker base. This will press the paper into a shape, which will then hold soil. Slide the paper pot off the handle. Set the newspaper pots into a seed tray and fill with growing media. Once plants root to the edge of the paper or grow through it, you can plant the seedling with the paper pot into the ground or into a larger pot. The newspaper will degrade naturally.

Another option is to skip the container altogether by using a soil blocker. A soil blocker is a metal form that creates compressed square soil plugs out of your growing media. These soil blocks sit in a solid seed tray, just like peat pellets or homemade seed plugs. It can be challenging to get lightweight soilless mixes to hold their shape, so you may need to add a binder in the form of denser organic matter or a bit of clay-based soil—or use a heavier potting mix. Of course, reintroducing bioactive matter into your soilless mix means it is no longer sterile. If you're an organic gardener, this may be more desirable, but you'll need to carefully manage moisture and humidity to minimize fungal and bacterial diseases.

As you become more experienced with propagation, you'll discover your favorite types of containers and which ones work best for you.

You can stamp out blocks of compressed potting soil to create your own plugs. Gently press the soil block plug down into an empty solid tray, leaving about ½ inch between individual blocks. Seeds or cuttings will root outward to the edges of the block, then you can transplant them. Water soil blocks carefully so you don't break them apart. >

GROWING MEDIA

〜

When buying or mixing your own potting soil or soilless mix for seedlings, keep in mind that you need to create a balance between water retention and good drainage. Small seedlings can dry out quickly and die, so they need growing media that can hold adequate moisture, but they are also susceptible to many soil-borne fungal diseases brought on by excess moisture. Vegetative cuttings can often tolerate soil mixes that are a bit heavier or rougher in texture. There are many types of pre-mixed potting soils available. Select a blend specifically appropriate for seeds, larger indoor plants, or outdoor plants, each of which has different characteristics.

The more organic matter in the growing media, the more opportunity for diseases and pathogens to become a problem. It can be beneficial to start your seeds and cuttings in a sterile, soilless growing media that contains coir, peat moss, biochar, or sphagnum moss mixed with vermiculite and horticultural perlite. You can even mix your own, using a combination of these primary ingredients. Mixing these ingredients creates a light, fluffy substrate that holds moisture but

allows enough air space for the delicate roots of tiny seedlings and cuttings.

Coir, or coco fiber, is excellent for growing seedlings and mixing into the growing media for larger plants. You can use coir alone or in combination with other ingredients, or even as a hydroponic substrate. It holds moisture well for an extended period while maintaining a porous structure that drains well. I love mixing coir with standard potting soil for containers, and even in outdoor garden soil, to help maintain a better soil moisture balance. Coir is considered a sustainably harvested product and renewable alternative to peat moss, because coconut trees can be grown anew.

Once hydrated, coir can hold onto moisture for an extended period of time. Store loose hydrated coir in a sealed container and you won't have to add any more water for many months. Some coir can be high in salts, which can interfere with the uptake of certain nutrients, such as calcium, so if you plan to use coir as a hydroponic substrate, look for brands that have been washed or composted. I also like to sprinkle a shallow

∧ Coir typically comes in a compressed block that you soak in water to rehydrate. After the coir soaks up water, it becomes light and fluffy in texture.

layer of coir on the soil surface when I start seeds as it helps keep the soil surface moist, aiding in seed germination.

Peat moss—a classic potting soil ingredient and hydroponic growing substrate—is harvested from living peat bogs, which are not replaceable, but peat bogs do keep making more peat. If we don't harvest more peat than the bogs produce naturally, peat moss can be considered sustainable.

Rice hulls, a byproduct of rice growing, are sometimes used in small quantities in growing media in order to aerate and loosen up the mix. They can also make the growing media more lightweight.

Biochar, which is created when organic materials such as wood, leaves, and manure burn in low-oxygen environments, can enrich the soil and also serve as a replacement or partial replacement for peat or coir. A 40/60 peat/biochar mixture has also been shown to make a good soilless mix for growing plants in hydroponic setups.

Many conventional, soilless seed-starting mixes are pre-charged with a granular synthetic fertilizer. If you are concerned about consuming foods grown with chemicals and prefer more organic gardening practices, use a bioactive soil instead. Look for a lightweight, soil-based seed-starting mix that contains organic matter and is labeled organic. These mixes are often premixed with natural organic components that will be broken down into fertilizer over time (such as worm castings, seaweed, and humus). They are a bit heavier than the soilless mixes, and you may experience more variation between dryness and wetness. There is also an increased risk of soil-borne diseases.

Over time, as you experiment with different types of soils and potting mixes, you'll find your favorites. You may even prefer to create your own custom blend. There are many recipes available in books and online. Different plants do well in different types of mixes so remember to take into consideration the plants you'll be growing. Dry garden plants such as succulents and cactus always need a loose, well-draining soil. Water-loving plants can be grown in a heavier mix. As a general rule, when starting seedlings, use a lighter mix that compacts and holds a bit more moisture. Once you move rooted plants to larger containers you can generally use chunkier mixes.

∧ Peat moss can be used alone as an inert growing media or mixed with a variety of potting mixes to loosen the mix.

∧ A loose, lightweight, general potting mix, good for all sorts of indoor plants, as well as transplanting young edible seedlings.

Growing Media Recipes

GROWING MIX FOR SEEDS

4 parts fine-screened organic compost
2 parts peat, coir, or biochar, moistened
1 part perlite
1 part vermiculite

GROWING MIX FOR CUTTINGS

1 part coir or peat
1 part perlite
3 parts coarse vermiculite

GROWING MIX FOR SUCCULENT SEEDS AND CUTTINGS

1 part builder's sand
1 part coir or peat

If you don't want to mix your own, look for packaged growing media labeled specifically for starting seeds or growing succulents. You can find plenty of conventional and sustainable options on the market, and you can always add some extra sand or other coarse material to loosen the potting mix. As your plants mature and grow larger, you will transition to mixes that can be heavier and provide better drainage. Most succulents and cacti will need a very loose mix with good drainage and aeration. For these types of plants, I like to mix in some decomposed granite.

Some plants, when taken as vegetative cuttings, may resist rooting in water and will do better rooted in an organic soil or soilless mix. Woody plants may rot before they root if submerged in water. Such cuttings may need additional help with rooting hormone and a heat mat. Remember, however, that some potting mixes hold more moisture or dry more quickly, depending on their ingredients. Many cuttings—often those resistant to rooting in water—will succumb to fungal or bacterial diseases faster when rooted in organic soils. If your plant roots quickly when placed directly in water, then the same plant may root just fine in organic soil. However, if your cuttings take a long time to root and are resistant to rooting in water, then you may need to switch to a soilless mix or inert media.

If you're potting up orchid offsets or divisions, be aware that you'll need an orchid mix rather than a standard potting soil. Orchid mixes are very chunky and loose, consisting of tree bark and other materials such as perlite, coir chips, and peat or sphagnum moss. Some mixes also contain charcoal or biochar. Most of the orchids you'll grow are epiphytes (air plants), meaning their roots grow exposed to the air, not directly in soil. Loose orchid mixes allow structure for the roots, but also plenty of air space and drainage.

∧ You can mix decomposed granite with other growing media for plants such as succulents and cacti.

∧ A typical chunky orchid potting mix.

ALTERNATIVE GROWING MEDIA

You don't have to use loose potting soil to grow your seeds or cuttings. There are many options for pre-formed plugs, both natural and synthetic, that can make propagation easy and tidy, while providing a good environment for your cuttings and seeds. Bioactive soils may contain pathogens that can rot the stem of your cutting before it has a chance to produce roots. If you're struggling to start seeds or root cuttings in potting soil, try switching to an alternative growing media to reduce fungal and bacterial disease issues. Coir and peat pellets, rockwool, foam root plugs, Oasis plugs, and hydroton are all common options. These inert substrates help balance the moisture-to-air ratio available to the cuttings and reduce pathogens.

Compressed soilless **pellet plugs**, also known as seed plugs, are compressed dry disks of peat or coir, plus a small amount of fertilizer, wrapped in a biodegradable film. Place the pellets in a seedling tray, cover them with water, and the pellets will expand. Set the expanded pellets in a solid seed tray (no need for an additional container), drop your seeds into the opening at the top of the film, and keep moist. Seed plugs are often sold with small seed-germinating greenhouse kits or you can buy them separately.

Be aware that the film that encases these peat or coir plugs, while biodegradable, can take a long time to break down. You can speed this up by using a sharp knife to make some slits in the film before transplanting your seedlings into the ground or larger containers. You can also create your own plugs using cheesecloth and coir or peat.

∧ Dry compressed coir pellet plugs in a watertight seed-starting tray.

∧ I set this tray of compressed coir pellets in my kitchen sink and covered them with water until they fully expanded. Now they are ready for a variety of seeds or cuttings.

Rockwool is a substrate made from the natural ingredients basalt rock and chalk; the mixture is melted into lava and then spun into a lightweight, felt-like substance. Rockwool is typically available in sheets or cubes, often with a small hole already punched for you to drop in your seed or cutting. You can also find kits that contain a water tray and sheet of rockwool or rockwool plugs. I find that rockwool is particularly effective for rooting cuttings, as it holds on to an even level of moisture.

Root plugs are typically made from hydrophilic, or water-absorbent, polyurethane foam. Biodegradable organic root plugs are also available. Root plugs are often inoculated with beneficial microbes to help seeds and cuttings get off to a great start. You can use root plugs to start seedlings that remain in a hydroponic setup or use organic root plugs to germinate seeds or to root cuttings. I love using kits that include a floating foam support for root plugs, specifically for cuttings, as it helps maintain consistent moisture, so your cuttings don't dry out.

Oasis is another type of inert foam material that can be purchased as sheets, blocks, or preformed plugs. You can start seeds or stick cuttings in Oasis

∧ A rockwool sheet in a watertight tray. Rockwool sheets often come with preformed holes or blocks for seeds and cuttings.

plugs. This type of foam has no impact on pH and, because there's no organic matter, disease is reduced. You can also purchase kits complete with a water tray and Oasis plugs. If you struggle to keep seeds moist enough for uniform germination, you might try Oasis plugs, as they are excellent at keeping seeds evenly moist for quick germination.

∧ Broccoli seeds germinating in an Oasis plug tray. Once rooted, these plugs will pop right out for you to transplant into a larger container.

< Foam root plugs sit inside a Styrofoam tray that floats in water, inside a watertight tray. Here you can see roots emerging from a citrus cutting. There are also pepper plant cuttings growing in the same tray.

Common Growing Media Ingredients

COIR A lightweight material made from the byproduct of coconut husks. Absorbs and holds water, while aerating the soil. Good for mixing with standard potting mix and garden soil, or topdressing when germinating seeds.

OASIS An inert growing media made from foam. It has no impact on pH. Good for rooting cuttings and germinating seeds.

PEAT MOSS Mined from peat bogs in Canada, peat moss improves water retention in potting soils.

PERLITE A volcanic glass heated to form small white balls. Mixed with potting soil to lighten the mix and improve drainage and aeration.

RICE HULLS A byproduct of the rice growing industry, rice hulls are very lightweight and can be mixed with soil or growing media to lighten the mix and improve drainage.

ROCKWOOL Rockwool is an inert media made from chalk and basalt rock. It is often used in hydroponic systems to provide structure for roots and hold water. Good for rooting cuttings.

SEED/PROPAGATION POTTING MIX A soilless, inert growing media made from a mix of products—vermiculite, perlite, peat—can be used for most seeding and cuttings.

VERMICULITE A lightweight material made from hydrated magnesium dialuminium iron silicate. Used to lighten seed-starting mixes or as a topdressing when germinating seeds.

If you're still struggling with cuttings in rockwool, or other inert substrates, you might want to dechlorinate your water, and slightly adjust the pH. Rockwool tends to have a pH of 7.0, which is a bit high for many seeds that are germinating or cuttings that are forming new roots. You can pre-soak the rooting media with dechlorinated or distilled water. Remember that you can naturally dechlorinate water by leaving it in a clear container in natural light for 24 hours. However, your municipal water may contain other chlorinating products that will take longer to break down. You can add a dechlorinator to your water to speed up the process and then add some citric acid or the pH down solution of your choice to adjust the pH a bit closer to 6.0.

If you are growing plants in a hydroponic system, your containers might come with a substrate called hydroton. These expanded clay pellets are often used as a supportive substrate in hydroponic and aeroponic growing systems. You can sprinkle seeds onto the hydroton where they will germinate, if humidity levels are adequate. Take note that tiny seeds may slip right through the hydroton, so if you're growing plants with small seeds, you may need to grow your transplant in a rockwool, Oasis, or foam root plug before transferring it to the hydroton-filled pots in your system.

∧ Clay hydroton pellets are often used to support plants in hydroponic grow systems.

ROOTING HORMONES

The trick with taking cuttings, just as with germinating seeds, is to get the cutting to root before it rots and dies. The sooner the rooting occurs, the better the chance for the cutting to grow successfully. While plants produce their own rooting hormones after being cut, some take longer than others. You can speed up the process, especially for difficult-to-propagate plants, by using rooting hormones on stem cuttings, leaf cuttings, and root cuttings.

The main ingredients in most rooting hormone products are auxins, which stimulate and aid in plant root growth. IAA, or indole-3-acetic acid, is the natural auxin found in plants that is most important for root development. Most rooting hormone products you buy will contain a synthetic form of IAA called IBA (indole-butyric acid) or NAA (naphthaleneacetic acid).

Apply these compounds to the base of your cutting to encourage faster and more vigorous root development. Rooting hormones are available in powder, liquid, and gel form, and you can purchase synthetic or organic products.

Rooting powder has a long shelf life, so it's easy to keep and store. To use a rooting hormone powder mix, dip the base of your cutting into the rooting hormone, making sure to coat the entire tip, and place it in your rooting media or substrate. To make more powder stick to the cutting, you can wet the end of the cutting first, then dip it in the powder.

If using a liquid rooting hormone, mix per instructions on the label (as some will be concentrated and must be mixed with water) and then dip the cutting in the solution, or allow cuttings to soak in the solution per label instructions. You can also directly add liquid rooting hormone solutions into the water reservoir of your aeroponic or hydroponic propagator—just make sure to follow mixing instructions.

Gels are the best option when water rooting as they are more likely to stay put on the cutting longer than a powder or liquid. When using a gel rooting hormone, you'll dip the end of the cutting into the gel, which will stick to the cutting. Only dip your cutting into the rooting hormone as deep as you plan to stick

The citus cutting on the left was not treated with rooting hormone. The cutting on the right, taken at the same time, was treated with rooting hormone. The cutting on the left will likely develop roots eventually, but it will take longer.

Wet the tip of your cutting, then dip into rooting powder.

it into the rooting substrate. Be careful not to rub or knock off all the rooting hormone as you place your cutting into its container.

Take care not to dip the cutting directly into the original rooting hormone container; instead, put some of the powder, liquid, or gel into another container, then dip your cuttings. If you dip your cutting directly into the original container, you can leave behind moisture or other organic matter than can cause the rooting hormone to break down, or molds and diseases to grow—especially in organic rooting hormones.

There are other natural rooting stimulants you can use if you don't want to buy a synthetic product or rooting hormone. Seaweed extract contains trace amounts of rooting hormones, plus additional important nutrients. You can use it as a natural rooting hormone and to reduce transplant shock. Directly water seaweed extract mixes into the containers that hold your cuttings or into potted plants to spur new root growth. Honey is another natural product that works as an antiseptic and antifungal product that can help cut down on pathogens—meaning honey helps prevent rot so your cutting can root. Simply dip the cutting in honey, just as you would with a gel rooting hormone. Willow extract is another highly effective natural root stimulator because it contains two auxin hormones: IBA and SA (salicylic acid).

Put some rooting gel into another small dish, then dip the end of your cutting into the rooting gel. >

∧ Some humidity domes come in taller sizes with a vent you can open and close to conserve or vent moisture and heat. This makes moisture management easier, allowing you to grow the seedlings under cover for longer.

Trays of microgreens are covered with humidity domes as they germinate on a kitchen countertop. >

HUMIDITY DOMES

Managing temperature and humidity can be tricky inside a typical home that runs air conditioners and heaters. The low humidity in our homes can quickly dry out young plants or seeds as they are trying to germinate. Don't have a greenhouse? No problem. By using a humidity dome on top of your seed trays, you can replicate greenhouse conditions on a smaller scale. You can purchase humidity domes separately to fit on plug trays or to cover other types of containers. There are also numerous mini-greenhouse kits available for purchase that come with everything you need to manage humidity for your young seedlings and rooted cuttings.

Many humidity domes have top or side vents. After the seeds have germinated and begin growing or cuttings begin to root, you can open the vents to slowly reduce humidity. If too much water begins condensing inside the dome, or you keep the humidity dome on too long, fungal diseases or rot could set in. While young seedlings and cuttings do need to remain moist, too much moisture in the air around the seedlings can promote damping off disease, which causes seedlings or cuttings to rot at the soil line and topple. Too much humidity also encourages powdery mildew on young seedlings, cuttings, and transplants. You can remove

⋀ Small plug trays with humidity domes, such as this, are perfect for starting cacti and succulent seeds.

the humidity dome or other cover once all your seedlings have germinated or cuttings have rooted and begin to produce new green growth.

You can also use a variety of other plastic containers with clear covers. You know all those take-out containers that come with clear plastic lids or small vegetable containers from restaurants or the grocery store? I use them frequently as mini-seed-germination greenhouses. It's a great way to recycle and reuse this type of plastic.

∧ I like to repurpose these small tomato containers as mini-terrariums for starting microgreens, a few seeds, or individual cuttings.

If you don't have a humidity dome or other type of plastic container, you can use clear plastic bags to enclose pots or small trays of cuttings to increase humidity. Choose a clear, non-porous, plastic bag large enough to accommodate your container and cutting without compressing them. Set an individual pot, several pots, or small seed trays inside the bag. Blow some air into the bag to inflate it, and then use a twist tie to close the end of the bag. This method is good when you're rooting individual or small groups of vegetative cuttings. You can also use bell jars and terrarium covers to increase humidity around vegetative cuttings and seedlings.

∧ A plastic bag covers this corn plant (*Dracaena* spp.) cane cutting.

⟨ This glass bell jar covers cuttings to maintain higher humidity.

A heat mat warming up the soil for new seedlings to germinate. >

CAPILLARY MATS AND HEAT MATS

Soggy seed trays can drown your small seedlings. To keep adequate and consistent moisture at the root level for your seedlings without drowning them, a capillary mat is a helpful tool. These mats resemble felt and are placed in the bottom of a solid seed tray. Put your seeds, plugs, or pots on top of the capillary mat and add just enough water to saturate the mat without any additional standing water. Capillary mats are also handy for directly growing microgreens.

Seedling heat mats can help speed up the germination process and ensure that seeds and cuttings root as quickly as possible, before they rot. This is important for crops that need warmer soil temperatures, especially when starting them indoors during colder months. Heat mats, placed directly under your seed tray (or rooting propagator), can warm up soil and water temperature by an average of 15 to 20 degrees F above room temperature. Most homes will have an average temperature in the 68°F to 74°F (20–23.3°C) range, but frequently are cooler in winter months. If you're propagating in your garage or basement in cooler months, the air temperature might be too cold for your new starts. Heat mats can help.

∧ Capillary mats typically come in the size of a standard propagation tray, but you can cut them to fit any container.

∧ I've cut a piece of capillary mat to sit in the bottom of a recycled plastic container, to wick moisture for pots of succulent seeds.

Be sure to use a heat mat specifically intended for seed germination. Seedling heat mats are designed to be moisture resistant and are insulated so you can place them on normal surfaces under your seed trays. (Do not try to recycle a heating pad designed to soothe your aching back, as it will present a fire hazard when used the wrong way.) Heat mats come in different sizes and you can purchase mats that link together to accommodate multiple seed trays. Once your seeds have germinated and begin growing, you can remove the heat mat.

If your growing space is a bit cooler than you'd like, simply keep the heat mat in place as the plants mature.

If you want to keep a close watch on temperature, you can add a timer or thermostat to your heat mat. A thermostat will allow you to set a specific soil temperature for your seeds, depending on what you're growing. The thermostat will also lower the temperature of the heat mat or even turn it off when the heat in your home matches the desired set point. That way your seedlings don't fry.

AUTOMATIC PROPAGATORS

~

If you plan on propagating plants on an ongoing basis or want to propagate plant species that are a little more challenging, you can invest in a propagator (also called a cloner, cloning machine, or propagation incubator), which can make vegetative propagation much easier. Propagators typically use hydroponic or aeroponic methods for rooting cuttings. You place your cutting through a small foam plug with a slit in it, then the base of the cutting is suspended in air, bathed in a continual mist or stream of water from inside the reservoir. These propagator setups will provide moisture, air, and rooting hormones at the right levels to achieve a high rate of success with your cuttings.

There are simple propagators that come with a small water pump and one rotating spray head to keep the base of your cuttings moist. Most will fit right on your kitchen countertop, are affordable, and very easy to use. Other systems get a bit more complicated, using larger pumps, aerators to add more oxygen to the water, and small misting nozzles to deliver a finer spray of water continuously to your cuttings. You'll need a little more space (and budget) for these larger propagators. The scale of your system should match your ongoing propagating needs. I've had much success rooting cuttings of a variety of plants in both simple, small propagators, and the large units.

You can add liquid rooting stimulators, nutrients, pH adjusters, dechlorinators, cleaners, and any number of products directly to the water stored in your propagator. Most products are clearly labeled as to when and how much you should add to your propagator water reservoir.

You'll need to balance humidity needs if you are rooting different types of plants together in the same propagator. For example, if you are pairing tomato cuttings with citrus cuttings, be aware that the tomatoes are more susceptible to quick fungal rot. You'll want to vent the humidity dome to discourage disease growth on the tomatoes, then remove the humidity dome once they began to root. Some cuttings will not need a humidity dome, but others will require one continuously until sufficient root growth has occurred. As you experiment with different cuttings, you'll learn which plants need more humidity than others.

Larger cloners use more powerful water pumps with an aerator and deliver a fine mist to the base of cuttings.

Small countertop propagators with nine to twelve plugs are an easy and inexpensive option for automated propagation. Look for units with humidity domes. >

∧ These Meyer lemon cuttings have been placed in a cloner under moderate levels of light. After 3 to 4 weeks, roots start to develop.

GROW LIGHTING 101

~

If you're growing in a space with little to no natural light, then artificial grow lighting is mandatory for plant propagation and growth. Seedlings, in particular, will almost always require grow lights. The time of year you propagate indoors will also influence your lighting needs. If you start your seeds or take cuttings indoors during winter months, you'll find that the lack of light and briefer duration of light can really slow things down. Short winter day lengths, a lower angle of the sun, and cloudy days all reduce the amount of natural light in your home available to plants.

Seedlings require intense amounts of light for specific durations; most windowsills simply are not bright enough, for a long-enough duration, for healthy seedlings. Conversely, vegetative cuttings you are rooting in soil, water, or other growing media don't require a lot of bright light—they only need the amount of light they'd receive in a bright windowsill. Cuttings can benefit from 24 hours of lower-intensity light until they root. So, the rule of thumb is that you need grow lights for seedlings and bright ambient light for cuttings.

DURATION AND DISTANCE OF LIGHT

Once a seedling shoot emerges from beneath the soil, both the amount of bright light and the duration of light are important. After sprouting, seedlings will need 14 to 16 hours of bright light to grow strong and sturdy. This is not because all seedlings are "long-day plants" (which is a reference to a photoperiodic requirement) as you might find them erroneously described in any number of online resources. It is simply because they need a lot of light and it takes that many hours of bright light to deliver what seedlings need.

Most indoor rooms, even those with large windows or plenty of diffused light, are not bright enough for baby seedlings. If your seedlings are elongating and stretching, they aren't getting enough light. Stretched seedlings will frequently topple and fail. If an overstretched seedling does survive long enough for you to transplant it into a larger container, the plant often has a tough time thriving and may not produce the desired results or harvest.

CAUTION
T5 HO MAX 2
ATTENTION
T5 H

Young tomato
transplants growing
under fluorescent
grow lamps.

TOMATO
ORANGE BLOSS
LYCOPERSICO

HARRY'S
SELECT

Vegetative cuttings, on the other hand, can be grown under continuous, less intense light until they develop roots. Bright light levels can burn your cuttings or cause leaves to yellow and drop. Typically, ambient light from a bright window is strong enough for unrooted cuttings and those that have not yet put on any new growth. Indirect light from grow lamps is also ideal. You can expose cuttings to 24 hours of light to speed up the rooting process. If you aren't using grow lamps and your cuttings are receiving only shorter durations of light from a bright windowsill, that's okay. Just know that it will take longer for the cuttings to root than if they received 24 hours of light. Once the cuttings have rooted and start pushing out new growth, you can place them under brighter grow lighting for that type of plant for the correct duration.

The closer your grow lamp is to your plants, the greater amount of light (and heat) will be delivered. As you increase the distance between the lamp and plant, less light (and heat) will be delivered. As stated several times already (I really mean it) seedlings need a lot of light. The light source must be very close to your seedlings as they germinate, as near as 3 to 4 inches from the seeds as they emerge from the soil. If the light is further away, less light will be delivered, and young sprouts can quickly stretch beyond their ability to remain intact.

Ratcheted cords or chains that allow you to move light fixtures up and down will enable you to place your lamps very close to seedlings as they are just emerging, then lift the light source higher as they grow. If you are using a fixture with adjustable shelves, you can create short and tall sections. Start the seeds on a short shelf that places them closer to the light, then move them to taller shelf sections as they grow—leaving more space between the plants and the light. Another method is

∧ Assorted cuttings water rooting in my kitchen windowsill

to prop up your seed trays with inverted plant trays or other containers to set them closer to the light as they are germinating; then remove the props once seedlings are larger and need to be set further away from the grow lamps.

As vegetative cuttings need lower intensity light until they root, you should place grow lighting at least 2 feet above your cuttings—higher is better. Indirect light from a grow light in the middle of the room or nearby is usually adequate. If you place grow lamps too close to your cuttings, the higher amount of light delivered can damage them. Once your cuttings are rooted and begin growing new leaves, then you can intensify the light levels and place them closer to the light source.

∧ These tiny microgreen seedlings have germinated successfully, but they are too far from the grow lights above and they will quickly begin to stretch if not moved much closer to the grow lamp. This distance of light, however, is perfect for rooting vegetative cuttings.

< Right now, these transplanted tomato seedlings are at a good distance from the grow lamps. As they continue to grow, I'll need to create more distance between them and the grow lamp.

∧ I like to create grow shelves for my seedlings and cuttings with an adjustable shelving unit. I make shelves different heights for different size plants, so I can move them around as they grow. I use large 4-foot fluorescent light fixtures (which can hold both HO T5 fluorescents and T5 LEDs) on these shelves in my garage, where there is no ambient light available. If you have wire shelving in your garage, be sure to purchase plastic shelf covers so they will contain any water drainage.

Sometimes I simply prop up seed-germinating containers, so they are closer to the light and then remove the props as they grow. I often have a variety of different seeds and cuttings growing on the same shelf, simultaneously. I can use smaller grow light fixtures in this space, which also has some ambient light from an adjacent window. ❯

If you are rooting vegetative cuttings, but you don't have the space or ability to move your lights up and down, you can start by removing two or three of the LED or fluorescent lamps from the grow light fixture to reduce light levels. As cuttings root and grow, you can reinsert the removed lamps.

Supplemental lighting may need to be turned on and off at times that aren't convenient for your human schedule. It is very easy to forget turning plant lights on and off at the right times for your plants. The best solution is to use simple or programmable timers. You can use the same simple timers you use to automatically turn on and off your house lamp while you're away on vacation or you can purchase digital timers that allow you to program a specific schedule.

∧ I removed three of the four lamps in this HO T5 fluorescent grow light fixture for my vegetative cuttings, leaving the light on for 24 hours a day until cuttings root. After that, I'll add lamps back to the fixture and reduce day length to 12 to 14 hours. You can see that I've used seed trays to hold my cuttings and covered them with humidity domes as they root.

< There are many simple timers available you can use to control plant grow lights as well as heat mats.

This CFL grow lamp emits light in the 6400K spectrum, meaning it includes more blue than red light, pushing it to the cooler end of the light spectrum. This grow lamp is appropriate for foliage plants and starting seeds and cuttings.

TYPES OF LIGHT

Without getting too deep into light science, it's good to know that the different spectrums (colors) of light will influence different types of plant growth and development. To speak very generally, blue (cool) light favors vegetative growth, while red (warm) light favors flowering. Full spectrum light on the cool side of the light spectrum is best for starting seeds and cuttings. Look for grow lamps labeled specifically for leafy vegetative growth; lamps in the 5000 to 7000K measurement range are appropriate. While your plants will still root and grow with more red light, or a warmer spectrum of light, you'll find that too much red light will cause them to elongate and stretch—not good for seedlings.

Full spectrum grow lamps will emit all colors of light in varying amounts, providing a mostly white or slightly tinted (warm or cool) color of light. Dual- or multi-band LEDs will generate a pink- or purple-colored light because they blend only red and blue light together, or sometimes red, blue, and one or two other spectrums of light. As red and blue light is used most efficiently by plants for photosynthesis, you can use dual-band LEDs that only mix the red and blue spectrum but be prepared for your space to be awash in pink light.

White light emitted from full spectrum LEDs versus pink light emitted from the same fixture, using only a dual-band red/blue light setting. >

GROW LAMPS

There are many different types of grow lamps to choose from and picking the right one can be a daunting task. Some lamps generate more light or more heat than others. Luckily for propagation purposes, you can rely on many of the smaller, less expensive types of grow lamps to meet your needs.

Skip incandescent bulbs. They don't generate enough light for your plants and they generate too much heat.

Cool spectrum high-output fluorescent lamps (HO T5 fluorescent lamps), CFLs (compact fluorescent lamps), and LEDs are typically the easiest and best options for starting young seedlings and rooting cuttings, as they generate relatively low levels of heat and can be placed close to your tiny plants. HO T5 fluorescents remain an efficient and inexpensive grow lighting option for the home gardener. LEDs can cost a bit more but offer the added benefit of lower heat output and longer lamp life. LED grow lamp technology is advancing quickly and there are many new options available.

If you are using HO T5 florescent lamps or LED bars for starting seeds, you'll typically use a fixture that holds two to four tubes or bars (lamps), depending on how many seed or cutting trays you have. Your goal is to deliver even amounts of light to the entire seed tray—even the edges. Small spotlight grow lamps (CFLs or LEDs) won't provide enough light for your entire tray of seedlings. If you are using an LED panel, then it needs to be large enough so light emitted reaches all your seedlings or cuttings evenly.

If you want to use a smaller light fixture and diffuse light over a larger area, you'll need to use reflective wall covering or start your seeds and cuttings inside small grow tents.

I find that seed-starting kits that only come with one small HO T5 fluorescent or LED lamp often don't provide enough light, causing seedlings to stretch unless they are also next to a bright window and receive a good amount of natural light. However, these lower light setups are perfect for rooting vegetative cuttings.

When I am growing seedlings, I use HO T5 fluorescent or LED fixtures that hold up to four lamps. These setups will provide the intense light that young seedlings need. If you already have fluorescent fixtures and you want to use LED lamps, you can insert LED bars that are retrofitted for fluorescent fixtures, which can emit less heat.

CFL spot lamps are available in small sizes (60 W) that can be inserted into standard light fixtures and larger sizes (250 W) that can be inserted into a grow lamp hood with a light reflector. The small bulbs are good options for small groups of seedlings, vegetative cuttings, or an individual houseplant—you can use the larger lamps with reflectors for larger groups of plants or edible crops.

If you really want to dig deeper into light science, how to measure and manipulate light, and all the specific types of grow lamps, be sure to pick up my book *Gardening Under Lights: A Complete Guide for Indoor Growers*. It will also help you learn how to grow food and ornamental plants indoors year-round, by using grow lighting.

Larger CFL lamps in light reflecting hoods are excellent to use when propagating bigger groups of plants. >

< This 12- by 12-inch, flat, dual-band (red and blue), integrated LED panel is perfect for starting small groups of seeds or cuttings that fit within one square foot of space. It can be hung at different distances from your plants to create higher or lower intensity of lighting. Take note: you cannot replace burnt out diodes in these types of integrated LED fixtures, but they should last for a long time.

Look for HO T5 fluorescents or LEDs when you're shopping growing lamps; skip the standard shop lights.

∧ Spotlight LEDs are good for lighting individual plants, but they won't deliver enough light for a tray of seedlings.

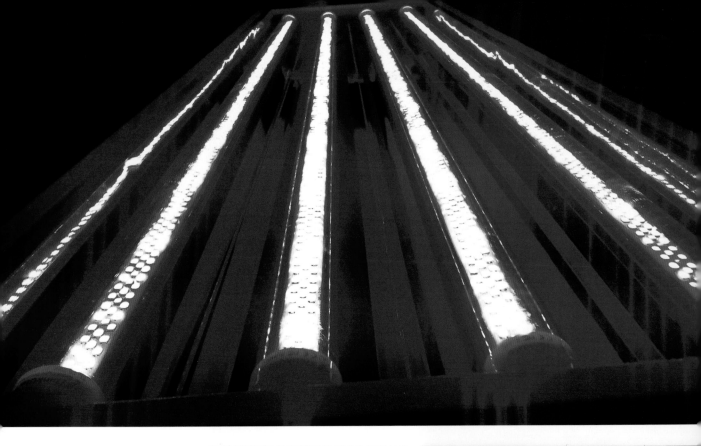

∧ These dual-band red and blue LED T5 bars fit into HO T5 fluorescent light fixtures and can be placed close to your seedlings.

This small CFL lamp fits in standard home light fixtures and can be used as a spotlight for individual cuttings or houseplants. ❯

∧ The same 12- by 12-inch LED panel can be used to light a larger area if all the light is contained and reflected inside a grow tent.

Seeds growing under 4-foot HO T5 fluorescent fixtures with four grow lamps. You can also replace these fluorescent tubes with LED HO T5 bars in the same fixture. Seedlings need this type of intense light.

Complete kits that come with one small fluorescent or LED lamp as the main or only source of light are perfect for rooting vegetative cuttings. However, if you're growing seedlings without much additional natural light, a single grow lamp won't provide enough light as seeds germinate. Kits with at least two grow lamps are better for seed starting.

Ripe seeds on an elephant garlic flower head.

STARTING

Your

SEEDS

How to Start Annuals, Perennials, and Edibles from Seed

A pea shoot emerging from the soil.

∧ An avocado seed sprouting new roots and shoot.

now that you have a handle on all your tools and materials, it's time to get down to the fun part! Once you catch the plant parenting bug, you'll discover that you never really have enough plants—and you need to make more. Nothing is more exciting than watching your new seeds sprout into baby seedlings or your cuttings take root. Propagation requires patience, but for true plant lovers, there's nothing more satisfying. Let's get to know a little more about how seeds work and how to become an expert germinator.

HOW SEEDS GERMINATE

Sowing new plants from seed is not only exciting, it's also an economical way to quickly grow a large quantity of plants. Plus, you'll be able to grow a much wider variety than if you only buy finished transplants from a garden center or online plant source. If you grow your plants outdoors, starting all your seedlings or cuttings inside can give you a big jump on the growing season.

Simply put, seeds are plants in embryo form. They contain enough energy to germinate and grow a new plant on their own once they are exposed to water and air, as well as the right light and temperature conditions. Until these conditions are met, seeds remain dormant, protected by a seed coat, called a testa, or chemical germination inhibitors. Seed coats prevent water from entering the seed embryo. Chemical inhibitors prevent the seed from germinating under the wrong environmental conditions. Otherwise, seeds would germinate at the wrong time, out of season, or in the wrong location and die. Inhibitors also make it possible for seeds to travel some distance away from the parent plant, which helps plant species spread geographically and improve genetic diversity.

∧ The seed coat of these horse chestnut seeds has been broken and both the radicle and hypocotyl have emerged. The cotyledons have not yet opened.

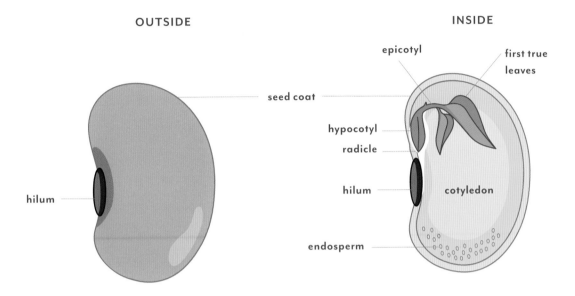

OUTSIDE

INSIDE

epicotyl

first true leaves

seed coat

hypocotyl

radicle

hilum

hilum

cotyledon

endosperm

hilum

∧ The anatomy of a bean seed. Bean seeds are protected by a seed coat, or testa.

Some seeds have a soft seed coat, while others have a very hard seed coat. Under the seed coat of some seeds, you'll find another layer called the endosperm, which serves as a food source for the seed. Under the endosperm, you'll find the seed embryo, which includes the radicle (root initial—the beginnings of new root tissue), the hypocotyl (shoot initial below the seed leaves), one or two cotyledons (seed leaves), and the epicotyl (the stem initial and true leaf initials that will grow above the cotyledon). You'll often see a small scar or eye on a dry seed. This small area, called the hilum, is where the seed was originally attached to the ovary on the mother plant.

Sometimes chemical germination inhibitors break down and a seed will germinate too early. Have you ever seen a tomato seed sprout *inside* a tomato fruit? We call this precocious germination, or vivipary. It happens when the hormones that regulate seed development degrade. Once the hormonal germination inhibitor is no longer present, the seeds will germinate, even under the wrong environmental conditions.

After the seed germinates and the radicle pushes into the soil or water, it will develop a primary root and the cotyledon (seed leaves) will open. Afterward, the first true leaf or leaves unfurl. Once you see the first true leaf emerging, you know your seedling is on its way.

The germination inhibitors inside this decaying tomato have broken down and the seeds inside have begun to germinate. By the way, don't eat those tiny tomato sprouts if you find them in your tomato fruit—tomatoes are in the nightshade family and the stems and leaves contain toxins. >

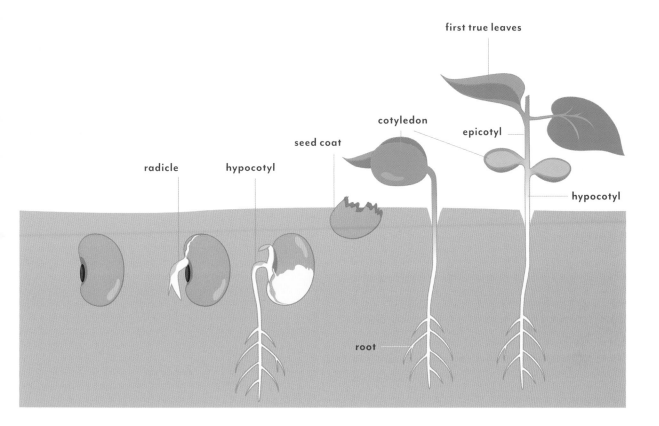

first true leaves

cotyledon

seed coat

epicotyl

radicle

hypocotyl

hypocotyl

root

∧ The process of seed germination. Once you see true leaves developing, your seedling will begin consuming more water, light, and nutrients.

PREPARING YOUR SEEDS

For most annual and edible seeds, normal germination occurs at optimal soil temperature and moisture levels without any special techniques. But you can speed up the germination process (or improve germination rates from older seed stock) if you pre-sprout them, a process called chitting or greensprouting. **Chitting** involves soaking the seeds, usually for 24 hours (some species require more time), before you sow them into pots or into the garden.

Seeds of stone fruits, such as peaches and apricots, nut trees, or citrus seeds, will root more successfully if you pre-sprout them. Chitting may also allow you to start seeds in cooler-than-optimal soil. In general, chitting is easiest with medium to large size seeds that you can handle easily, but you can do it with smaller seeds, too, if you're careful.

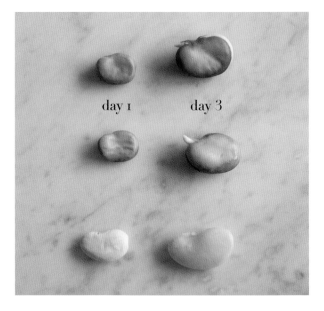

day 1   day 3

︿ As these seeds soak up water during chitting, they swell, and the germination process begins. Chitting certain types of seeds can speed up germination and improve germination rates.

Borage, an herb with beautiful blue edible flowers, is a prolific seed producer. It self-seeds in my garden, while leaving plenty of seeds on the plants for me to collect and save. ❯

How to Chit Seeds

1 Moisten some dish towels, paper towels, or newspaper to the dampness of a wrung-out sponge, then set the damp material in a tray or on a plate.

2 For seeds that sprout quickly, such as beans (1 to 3 days), simply spread your seed onto the moist surface. For seeds that take longer and need more constant moisture to sprout—and for larger, harder seeds that take longer to sprout, such as many succulents and cacti (several weeks)—insert the moistened dish towel or paper towel into a small plastic baggy, place the seeds inside the moistened towel, and seal the baggy.

3 Place in a warm spot in your home.

4 Seeds will absorb the moisture and swell, and some will germinate and begin to sprout. Immediately plant these sprouted seeds into a water rooter, growing media, or seed plugs.

5 If your seeds have been molding, then dilute a 1:25 ratio of hydrogen peroxide to water and wipe the seeds with the solution before you place them into the moist towel.

Some seeds are more difficult to germinate than others because they have a very hard, protective seed coat. Seeds with hard seed coats can typically be stored successfully for much longer than seeds with soft seed coats—but they also take longer to germinate. These are the types of seeds that can travel long distances over long periods of time in nature before they germinate, or even make their way through animal digestive tracts (where the seed coat is broken down) before they are deposited to germinate. Some seeds with hard seed coats must be exposed to fire to break their dormancy and germinate. These are all evolved, survival tactics for each species of plant.

If you're not sure if your seeds have a hard seed coat, give them a squeeze or a tap on a hard surface. A seed has a hard seed coat if no amount of squeezing or tapping produces any give. You may have discovered your seeds have a hard seed coat because they never germinated, or they took an extensive amount of time to germinate.

While most annual vegetable seeds do not require pre-soaking or any special preparation for germination, seeds of some natives, perennials, and fruits with hard coats will require a bit of extra work on your part, whether it be a longer chitting/soaking period, scarification, stratification, or inoculation with rhizobium bacteria. Information on the seed packet usually includes the type of seed coat and any special germination needs.

Seed scarification involves scraping away part of the hard coating to expose the seed to water and gases that trigger germination. In the natural environment, temperature, soil microbes, and even fire can break down seed coats. Animals eat seeds, which are then exposed to stomach acid which breaks down the seed coat.

< Certain succulents and cacti can be tough to germinate. Of the four types of succulents I seeded in this tray, only two seeds of one species (*Aloinopsis schoonee-sii*) successfully germinated. The seeds of the other species probably needed a different type of seed preparation to germinate.

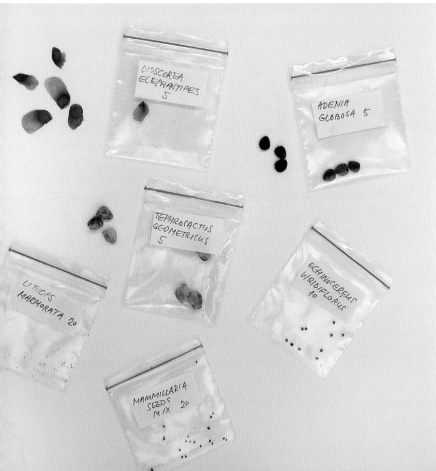

< Seeds of succulents and cactus can vary dramatically in size, shape, and hardness. Some are soft and barely bigger than a speck of dust (*Lithops* spp.), while others are large with hard seeds coats that require scarification (*Tephrocactus* spp.).

If you have had trouble getting certain types of seeds to germinate and they have a very hard outer coat, try scarifying them with one of these easy methods.

HOT WATER Boil water and pour it into a bowl. Let it cool for about 30 seconds, then add the seeds. Allow the water to cool to room temperature and the seeds to swell, then remove them. The amount of time needed for seeds to swell will depend on the plant variety. It may take a few hours or a couple of days. Soaking for 12 to 24 hours is recommended, but don't let seeds soak for more than 48 hours.

FILE Use a small file or sandpaper to rub away or nick a small section of the seed coat. (Do not rub off the entire coat.) Make your mark on the opposite side of the seed-eye, where the shoot will emerge, so you don't damage the eye.

SAND Roll or rub your seeds with sand or fine gravel for a few seconds. You can do this between sheets of paper or another flat surface. Don't crush the seeds, you only want to scrape away a bit of the seed's surface coating.

Some seeds do not have a hard seed coat and will instead go into a dormancy period that can make germination a long affair. You will have to wake these seeds from their slumber using a process called stratification. Many natives, wildflowers, herbaceous perennials, and grasses fall into this category.

Stratification involves creating the most favorable environmental conditions for the seeds—conditions in which they would germinate in nature. Seeds and plants that go into dormancy typically come to life after the return of spring temperatures and rainfall. You must mimic the cycle of dormant winter season followed by natural germination season. For starting seeds and growing indoors, create an artificial winter by chilling the seeds.

Some seeds experience a double dormancy, so they will not germinate or put on green growth above ground until they have experienced two winter seasons. Species in the lily family require a double dormancy. To speed up production, trick these species by using a repetitive stratification process.

There are two common methods for seed stratification: moist and dry.

MOIST STRATIFICATION Mix your seeds with damp sand and seal them in a container. Store the container in your refrigerator (not the freezer) for 1 to 2 months. Then remove the seeds and allow them to dry at room temperature. Using a seedling heat mat, sow and germinate the seeds per the instructions on the seed packet.

DRY STRATIFICATION This involves storing dry seeds at freezing temperatures for 1 month or more, depending on the species. Never put moistened seeds in the freezer, as this can kill them.

Seeds of many native wildflowers and prairie grasses will benefit from dry stratification for 1 to 3 months before you try to germinate them. Other native and woodland-type wildflowers perform best with moist stratification for about 1 month. Research the plant you want to germinate to determine whether it needs any chitting, scarification, or dry or moist stratification to germinate successfully.

Legume plants, such as beans and peas, have developed beneficial symbiotic relationships with several species of **soil-borne bacteria** (*Rhizobium* spp.).

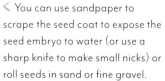
‹ You can use sandpaper to scrape the seed coat to expose the seed embryo to water (or use a sharp knife to make small nicks) or roll seeds in sand or fine gravel.

‹ Seeds of many types of wildflowers, such as these beautiful blue gentian, require or benefit from stratification. Without a couple of months of moist stratification, gentian seed are stubborn germinators.

∧ These pea seeds will benefit from soaking in, or being watered with, rhizobium bacteria.

∧ These lemon seeds are clones of the mother plant. You can plant these seeds and grow plants identical to the mother plant.

These rhizobium bacteria help form special nodules in the roots of legume plants that enable the plants to fix their own nitrogen from the atmosphere and soil, so that the plant can absorb it. Nitrogen fixation is the conversion of atmospheric nitrogen (N_2) into ammonia (NH_3), which is the form of nitrogen plants can absorb and use. Essentially, the bacteria help the plants make their own fertilizer. Seeds of legume plants can benefit from inoculation with rhizobia before planting. In fact, if you buy rhizobia from a garden center or online, it will probably be labeled as garden inoculant and will be in a powder form. You can mix it with water and pour it into the soil where you will be planting legumes outdoors or soak your legume seeds in the inoculant before planting. Follow label instructions.

Clone Seeds

Many varieties of citrus and mangoes produce what we call apomictic seeds, or clone seeds. These clone seeds develop without the need for male genes or pollination and allow you to grow new identical specimens of your mother plant from seed instead of a cutting. So, if you save seeds from your lemon tree, you can germinate them and consistently grow the exact same type of lemon.

GROWING TRANSPLANTS VS. DIRECT SEEDING

On the seed packet you will find information on when you can expect flowers or fruit from a mature plant, also known as days-to-harvest. If your tomato seed packet states 85 days, that means it typically takes 85 days to start harvesting fruit . . . but 85 days from when, exactly?

The number of days-to-harvest can vary significantly, depending on whether you direct seed a crop into the garden or first grow a transplant. Tropical crops, such as eggplant, peppers, and tomatoes, are typically started early indoors as transplants, allowing for bigger plants in the outdoor garden once temperatures are appropriate. For these plants, days-to-harvest is counted from the date they are transplanted into their final container or outdoor garden location. So, for the 85-day tomato variety, harvest time comes in 85 days plus the 8 to 10 weeks it took to grow a garden-ready transplant from seed—or 135 to 149 days (19 to 21 weeks) total to produce harvestable fruit.

∧ Before you sow seeds, be sure to refer to the seed packet for information on seed prep, timing, planting depth, spacing in the garden, and days-to-harvest.

When starting seedlings indoors for your outdoor garden, you must prepare the seeds early enough to plant outside at the optimal time. Crops such as

∧ I direct seeded this mixture of edibles and flowers—including calendula, kale, Swiss chard, borage, bush beans, and lettuces—into my vegetable garden. I sowed the seeds directly into the garden beds when outdoor temperatures were favorable, rather than starting transplants indoors.

broccoli, kale, and cabbage can be ready to plant outdoors in as few as 5 to 6 weeks from the date you germinate them. Leeks and celery, on the other hand, can take a full 12 weeks. Research how long it takes to grow a planting-ready transplant, and then count backward from the outdoor planting date for your area. That's when you must get your seeds started indoors. Make sure to transplant plants shortly after germination as they can become stunted if you allow them to live too long in a small container.

If you direct seed your crop into the garden or the container in which it will grow indoors, the days-to-harvest number on the seed packet is calculated from the date of germination. Plant these crops when indoor or outdoor temperatures are ideal. It's best to direct seed root crops (carrots, turnips, beets, radishes) and some large-seeded vegetable crops (beans, broad beans, squash, corn) straight into the garden or final growing container, as they won't always transplant well.

Other crops fall somewhere in the middle. You can direct seed lettuce and many other salad greens, for example, into their final container or garden spot, but you can also grow them as transplants first, then

∧ I direct seeded lettuce into large containers, where it will continue to grow under lights. I'll continue to grow and harvest this lettuce indoors.

∧ Root crops, such as carrots, do best when you direct seed them into your outside garden beds or large containers.

‹ Days-to-harvest varies, depending on whether you direct seed or transplant your edible crops.

bump them up into larger containers or set them in an outside garden.

Before planting seeds, be sure to thoroughly moisten the growing media, pellets, or plugs. Plant your seeds at the depth recommended on the packet. Typically, you can plant larger seeds deeper and smaller ones more shallowly. You should sow some seeds, such as lettuce, on top of the soil and lightly press them down. Mist the lettuce seeds to moisten them, but don't cover them with soil or they won't grow.

Always make sure that large containers you're direct seeding into have drainage holes. And for these larger containers, take care that you don't overwater them in the early stages of seed germination and growth. With more soil volume, the larger pots can stay soggy, leading to rotted seeds and seedlings.

You won't always achieve 100 percent germination from a group of seeds, whether you buy or harvest and keep for the next season. You can sow two seeds per container, cell, or plug, so if one seed doesn't germinate, you have a backup. If your seed is on the older side, you might want to sow three seeds per cell, to improve your odds.

MOISTURE AND HUMIDITY

Maintaining proper moisture is key to successful germination and healthy seedlings. The growing media should always be damp to the touch, like a wrung-out sponge. Never let it dry out, but don't let it stay soggy, either. Do not overwater seeds that haven't yet germinated, as they can rot if they stay too wet. Use a plant mister bottle to keep the soil surface moist until seeds germinate and start developing, then begin adding a little water to the seed tray so the growing media can absorb it as seedlings mature. While succulents and cacti will need you to dial back on watering once they begin to mature, most will need similar moisture conditions as many of your other annual and perennial seeds when germinating—so don't let them dry out too much when they are just getting started.

One trick to maintaining even moisture as seeds germinate and to keep the surface of the soil from drying out is to sprinkle a thin layer of a material such as coir or vermiculite on top of the soil. I prefer to use coir. These materials will hold additional moisture, which can help prevent your seeds from drying out before they germinate.

To help maintain adequate moisture and humidity place a humidity dome, or other type of clear cover, on top of your seed trays or pots. Remove the cover once all your seedlings have germinated and begin to produce green growth. If you keep the dome on for too long, fungal diseases or rot could set in. While young seedlings need to remain moist, excess moisture can promote damping off disease, which causes seedlings to rot at the soil line and topple over. Too much humidity also encourages powdery mildew on young seedlings and transplants.

You can also use a felt-like capillary mat to keep seed plugs or small transplants from drying out or staying too soggy. Set your capillary mat under your seed plugs or pots and saturate it until it is moist to the touch. If you're growing microgreens, you can also germinate the seeds directly on a capillary mat.

Let's talk a bit more about damping off disease, as it can be an incredibly frustrating situation for new

‹ When sowing these pepper seeds, I placed two seeds per cell, approximately ½ inch beneath the soil surface.

‹ After seeds are sown and covered, mist the soil surface to gently saturate all the growing media.

Sprinkling coir, or vermiculite, on top of the soil surface after you place seeds will help balance moisture during seed germination. >

Pea seedlings emerging from the soil. >

^ Seeds germinating under humidity domes.

and experienced seed starters alike. Damping off is a disease that young seedlings rarely survive. Species of fungi (*Rhizoctonia* spp. and *Fusarium* spp.) in conjunction with the water mold *Pythium* spp., are the typical culprits that cause damping off disease.

While older seedlings and transplants can typically fend off such attacks by the fungi that cause damping off, germinating seeds and young seedlings just developing their first set of true leaves are particularly susceptible.

^ After germination and true leaf development, I like to use a small squirt bottle to gently water young seedlings and succulents right at the root zone, without getting water on the foliage.

Signs your seedlings are suffering from damping off

- Your seeds never emerge from the soil

- The cotyledons are discolored or look waterlogged or mushy

- The seedling stem becomes thin and water-soaked looking

- The new leaves wilt or look discolored

- You find no roots on your seedling or the roots are discolored and stunted

- White mold-like growth develops on seeds or seedlings in high humidity.

These microgreens have germinated on the capillary mat. I've left an area unseeded so you can see how they root directly into the mat. When you harvest, you'll simply cut away the microgreens at the base of the stalk using some sharp snips, and then compost the used capillary mat. >

∧ If you want to skip the soil, microgreen seeds will root directly into a capillary mat set inside a solid plant tray. Wet the mat first so it's damp, sow your seeds either in a solid or row pattern, then mist the seeds with a spray bottle. Place a humidity dome over the tray.

These microgreen seedlings succumbed to disease before they could develop any further. >

∧ Due to several less-than-ideal conditions, such as excess moisture and the wrong temperature range, many of these lettuce seeds have succumbed to mold growth before they could germinate.

Conditions were too wet and dark when these microgreen seedlings were beginning to germinate, and some mold began to grow on the seeds. I removed the humidity dome for a day or two and increased the light and was able to save this crop. >

Moisture management is crucial to keeping damping off disease at bay. I find that you'll have the most trouble with this disease in cold, damp soil, in low-light conditions, or in too much humidity. Using heat mats can warm up the soil to optimal temperatures and speed up germination; but if your emerging seedlings are not getting enough light or are kept covered by a humidity dome too long, damping off disease can still develop.

Another way to minimize chances of damping off is to use a sterile seed-starting mix. If you are reusing pots or other containers, be sure to sterilize them between plantings by soaking in a 10 percent bleach solution for 20 to 30 minutes. Also avoid adding fertilizer to your seeds before or right after they germinate. Fertilizers can encourage fungus and mold growth. Commercial growers will use synthetic chemicals to prevent damping off, but you probably don't want to do this in your home or for edible crops. A handy alternative is to water your seedlings with a hydrogen peroxide solution. Hydrogen peroxide oxygenates the soil, which kills off many fungi and bacteria. Mix 1 teaspoon of hydrogen peroxide to 2 cups of water and then use the solution to water or mist your seedlings.

You'll also need to take care with your watering habits for young seedlings and succulent plants. As seedlings grow, you're best to shift from a spray bottle—which wets the foliage—to small watering cans or squirt bottles, so you can deliver water directly to the root zone without getting water on the foliage. Keeping excess water off the foliage will help reduce fungal disease problems.

TEMPERATURE

Both air and soil temperature affect the speed and success rate of seed germination and growth. Each type of plant has a different optimal temperature range, based on its natural environment. Be sure to check the seed packet for specific optimal temperature ranges for germination. Many seeds germinate well in the 68°F to 80°F (20–26.6°C) range for both soil and air temperature. If temperatures are too cold or too warm, some seeds take a very long time to germinate or may not germinate at all.

Your goal is to get as many seeds to germinate as quickly as possible. For crops that need warmer soil temperatures, such as tomatoes, basil, or zinnias, especially when starting them indoors during colder months, a seedling heat mat can speed up the germination process and ensure success. Heat mats, which are

∧ Different types of seed will germinate faster than others. This 'Genovese' basil is clearly an overachiever, when compared to the slower tomato and pepper seeds sown at the same time.

∧ Small clip fans are handy for improving air circulation and cooling the air temperature around your grow lights.

placed directly under your seed tray, can warm soil to an average of 15 to 20 degrees F above room temperature. Once your seeds have germinated and begin growing, you can remove the heat mat. If your growing space is a bit cool, keep the heat mat in place as the plants mature. Remember that you can use a thermostat with your heat mat to turn it on and off automatically.

Seed packets typically provide an estimated germination time based on optimal temperature ranges outdoors, but you can typically shorten germination time in a controlled environment with heat mats.

If you are growing cold-weather lovers, such as spinach, indoors in the summer, you might run into problems with temperatures being too hot for successful seed germination. Sometimes the problem is the heat emitted from your grow lighting, or you may be growing in a garage or other part of your home that gets very warm seasonally. You can always position small fans near your grow lamps when you need your lamps close to your seedlings and the room temperature is already warm.

Ideal Seed Germination Temperatures for Popular Plants

COMMON NAME	IDEAL TEMPERATURE (°F/°C)	DAYS TO GERMINATION	SEED LIGHT EXPOSURE
basil	70/21.1	7	cover
bean	80/26.6	5	cover
broccoli	80/26.6	5	cover
carrot	80/26.6	7	light cover
cauliflower	80/26.6	4	cover
celosia	75/23.8	7	cover
corn	90/32.2	3	cover
cosmos	70/21.1	4	cover
cucumber	85/29.4	3	cover
dianthus	70/21.1	5	cover
eggplant	80/26.6	5	cover
flowering tobacco	75/23.8	10	no cover
four-o-clocks	70/21.1	5	cover
geranium	75/23.8	7	cover
gerbera daisy	70/21.1	7	light cover
globe amaranth	75/23.8	10	light cover
kale	70/21.1	4	cover
lettuce	70/21.1	3	no cover
marigold	80/26.6	4	cover
moss rose	80/26.6	10	no cover
mustard	80/26.6	3	cover
nasturtium	70/21.1	5	cover
parsley	85/29.4	12	cover
pea	75/23.8	6	cover
pepper	85/29.4	8	light cover
petunia	75/23.8	10	no cover
radish	85/29.4	3	cover
snapdragon	70/21.1	8	no cover
squash	85/29.4	3	cover

COMMON NAME	IDEAL TEMPERATURE (°F/°C)	DAYS TO GERMINATION	SEED LIGHT EXPOSURE
sunflower	75/23.8	3	cover
Swiss chard	85/29.4	5	cover
tomato	70/21.1	6	cover
vinca	70/21.1	14	cover
watermelon	90/32.2	3	cover
zinnia	75/23.8	3	cover

CULLING THE HERD

Now that your seeds have germinated, it is time to thin their numbers. This part can be tough—no one wants to kill the seedlings they just grew. Remember those two or three seeds you placed into the seed cell? Sometimes all of them germinate and you end up with a tiny seedling forest. When too many seeds germinate too closely together, the seedlings can struggle. It is tempting to let them all continue to grow, but your seedlings will be better off if you cull the weakest ones. More than one seedling per cell causes too much shading and resource competition, resulting in weaker seedlings overall.

After your seeds have sprouted, choose the strongest, stockiest seedling in each cell and snip the remaining seedlings at the base. Throw the excess seedlings on your salad or feed them to any critters that would appreciate some greens.

LIGHTING YOUR SEEDLINGS

Many seeds require darkness underneath the soil to germinate properly. When germination occurs in darkness, all the seed's energy funnels into root growth. Once the shoot breaks the soil surface and is exposed to light, things change dramatically. Root growth slows and shoot elongation accelerates.

Some seeds require exposure to light—and specific types of light—in order to germinate. Lettuce seeds, for example, need to be exposed to red light or they won't sprout. (Red light will be present in natural ambient light and full spectrum grow lighting.) It is a good practice to check the seed packet for information about germination light or darkness requirements before starting your seeds.

∧ Lettuce seeds need exposure to light to germinate.

Remember from Grow Lighting 101 (see page 76), once your seedlings germinate and emerge from the soil, they will need 14 to 16 hours of continuous bright light to grow strong and sturdy. If your seedlings are elongating and stretching, they aren't getting enough light. Place your grow-lighting fixture 3 to 4 inches above your seeds. Place a lighting timer on the fixture so your lamps turn on and off automatically at the right times of day and night. A 6:00 a.m. to 8:00 p.m. light schedule works well for bright grow lamps. If you are using lower-intensity lights, leave them on from 6:00 a.m. to 10:00 p.m.

After your seedlings have grown to have several true leaves, you can begin to raise the grow lamps, so your young plants don't burn.

∧ If multiple seeds germinate in the same cell, keep the strongest and snip off the extras.

FEEDING YOUR SEEDLINGS

Once your seedlings or cuttings have developed four or five true leaves, and are no longer pulling resources from the seed, it is time to start feeding them with fertilizer. Small seedlings cannot handle full-strength fertilizer, so be sure to dilute your product with water by one-quarter to one-half the recommended rate.

There are many different types of beneficial fertilizers to use on young seedlings, but it's best to use a liquid suspension rather than a granular formula. Liquid seaweed, fish emulsion, and liquid humus products are effective, natural options that won't burn young plants. Dilute these products to half strength when applying to small seedlings, or choose a liquid fertilizer labeled specifically for seedlings and follow the manufacturer's instructions. If you choose a synthetic fertilizer, you should dilute it by one-quarter of full strength, as these products are more likely to burn or damage your young plants than naturally derived fertilizers.

Feeding your seedlings once per week with diluted fertilizer is adequate, or every two weeks if you are using a stronger fertilizer mixture. As plants mature you can increase the rate of fertilizer application to full strength. Do not mix your fertilizer at a stronger-than-recommended rate. While there is little danger of damaging your seedlings with too much organic fertilizer—you'll merely be wasting product—a synthetic fertilizer overdose could burn your plants beyond repair.

∧ Chlorosis can appear as an overall yellow cast to the leaf or interveinal chlorosis (yellow between green veins), as shown in the photo, because of a lack of chlorophyll. Chlorosis signals potential nutrient deficiencies (iron, manganese, zinc). It can also signal poor drainage, soil compaction, or high soil or water pH.

HOW TO COLLECT AND STORE SEEDS

~

If you grow plants in your outdoor garden or in containers and allow them to flower and go to seed, you should be sure to collect and save some seed. Collecting and saving seeds from the plants you're already growing is a sustainable approach to your gardening endeavors and a smart way to save money.

There are two main types of seeds: wet and dry. Dry seeds develop inside a husk or a fruit pod that dries completely so seeds are visible. Beans, okra, peppers, onions, and herbs such as dill produce dry seeds.

Harvest dry seeds while still on the plant, once they have completely matured and dried. The dry pods will be easy to open or will burst open on their own, and the seeds inside will be hard and completely dry. Once you have harvested the seeds, leave them on a towel in a cool, dry place for a few days to ensure they are completely dry before you store them.

Wet seeds are produced from fleshy fruits such as tomatoes, eggplant, and squash. They typically remain inside a large amount of flesh and are not visible unless you break open the fruit. To harvest wet seeds, break open a mature fruit that has begun to shrivel, then separate the seeds from the fleshy part of the fruit. After the seeds are completely clean, spread them out on a dry surface for several days before storing.

Some wet seeds perform best if you ferment them before storing. Fermenting mimics the process the seeds go through on the vine as they mature and enables you to eliminate fungal activity that could damage the seeds or prevent germination in the future. Fermentation also breaks down germination inhibitors that may remain in the seed coat.

Seeds of fava beans, cucumbers, bush beans, and peas. Large seeds like these are easy to pre-soak and direct seed in the garden. >

I allowed these peppers to dry completely on the garden plants before harvesting the seeds within. >

This milkweed seed has matured on the plant and is ready to be collected or spread around my garden by the wind.

To ferment seeds, place them in a bowl of water with some of the fruit flesh remaining around them. Use about twice the volume of water as the amount of seed. Place the container in a warm area, where temperatures are between 75°F and 80°F (24–27°C), for 2 to 5 days. (Alternatively, look for bubbling or mold to appear on the surface of the water, which could occur sooner.) Once the fermentation is complete, the viable seeds will sink to the bottom of the container and the bad seeds and any remaining debris will float to the top. Do not let the seeds ferment too long or they can start to germinate. Spread out the good seeds on a dry surface for several days, then store them as you would dry seeds.

Place your dried seed in sealed containers—such as envelopes, paper seed packets, or jars—and store in a cool, dry, dark place. If you store seeds in an area where they will be exposed to very hot or very cold temperatures they can degrade.

Keeping seeds in a cool, dry place will help extend their life. Be aware, however, that the longer you store seeds, the more their germination rate decreases. For example, you may achieve 100 percent germination from seeds you've harvested and stored within the same year, but 3 years later you may get only 60 percent germination. These rates will differ depending on plant variety and the conditions in which seeds are stored.

As you get the hang of seed-starting basics, you'll develop confidence to branch out with propagating more difficult species. Experiment and have fun!

∧ Once tomatoes are large enough to develop mature seed, you can separate them from the fruit flesh.

∧ These fermenting tomato seeds are starting to bubble and drop to the bottom of the bowl. In another 1 or 2 days they'll be ready to dry and store.

∧ I use paper seed packets with my own labels, so I remember what I've stored and when it was harvested. Gifting seeds you collected, or trading them with your friends and family, is fun and rewarding.

< I seeded the same variety of lettuce in all these seed plugs . . . but the seeds were 3 or 4 years old. Only about 50 percent of the seeds germinated. The rest were no longer viable.

Spores

In non-flowering plants such as ferns, propagation from seeds is not possible because ferns don't flower or develop seeds. Instead, ferns make spores. These spores spread to moist areas, are fertilized, and generate new ferns. So, while you can't collect seeds from ferns, you can easily multiply ferns with spores, simple division, root cuttings, rhizome cuttings, and tissue culture.

∧ These tiny fern spores from a fern frond were dusted onto a cloth so they could be visible.

∧ A fertile frond from a holly fern. The small brown areas on the undersides of the leaves, called sori, contain spores.

How to Grow Ferns from Spores

1 Lay a fertile frond (with sori facing down) onto a sheet of paper. Spores will drop onto the paper within 24 hours if they are ripe.

2 Fill a shallow container with organic compost and add some earthworm castings. Dust the spores on top of the soil and cover with a clear plastic container or enclose in a clear plastic bag.

3 Place under a single cool fluorescent grow lamp for 12 to 14 hours a day. No direct hot sunlight.

4 In several weeks, a tiny green carpet will begin to develop, which are the prothalli of the ferns (they contain sperm and egg.) When they reach about ¼ inch, mist them with dechlorinated water. You should then see tiny sporelings (plantlets) developing.

5 Gently divide clumps of sporelings into plug trays filled with a lightweight indoor potting mix and cover with a clear plastic cover. Once fern plants have filled out the space, you can transplant them into 4-inch or larger containers.

WATER ROOTING

A Beginner's Guide to Windowsill Cuttings

Many plants, such as these assorted tropical houseplants and herbs, can be propagated simply by taking cuttings from the plant and rooting them in water.

Water rooting is often the gateway drug for those interested in making more plants. Most of you probably had a mother or grandmother with a collection of plant stems rooting in glass jars on the windowsill—with an avocado seed or two mixed in for fun. You may have gotten your own first try at plant propagation by water rooting some form of plant in school.

Lucky for you, many plant cuttings and some seeds don't need soil or potting mix to root and grow new plants; all they need is water. Large seeds, such as those of avocado or oak acorns, can be suspended partially in water where they will sprout new roots. You may want to pre-soak your large seeds before you suspend them for water rooting.

Vegetative cuttings of many types of plants can be rooted by directly submerging stems in water. I recommend using soft, fleshy stem cuttings and leaf cuttings, which we'll discuss in detail in the next chapter. Many tropical houseplants and fleshy garden perennials root easily in water. In fact, you can root many types of stem and leaf cuttings in both water and soil or other growing media. Depending on the type of plant, water rooting can often be faster than soil rooting, and it is more fun to watch.

Typically, woody and semi-woody plant material is harder to root in standing water, as it will often rot before it roots. You will usually have more success with woody or tough stem cuttings (i.e., citrus, gardenia, roses, other shrubs and trees) by using an organic or synthetic growing media, or an aeroponic propagator (cloner).

∧ Who can resist rooting an avocado seed now and then?

All you need is a windowsill to light most of your water-rooted cuttings. Until they develop roots and are ready to be transplanted, low-light conditions are just fine for these vegetative cuttings. Once the cuttings have developed roots and are ready to be transplanted, you'll need to move the plant to the type of light conditions it traditionally favors. If you're sprouting large seeds in water, such as avocados, remember that seedlings need brighter light than vegetative cuttings. Be sure to place emerging avocado seedlings in a bright window or use a grow light.

A piece of my monstera philodendron broke away from the mother plant. Because it included a node with the beginnings of an aerial root, I was able to place it in this lovely water pitcher to root. After a couple of weeks, new roots emerged, and the cutting was ready to be potted up.

∧ This vintage test tube set is perfect for water rooting an assortment of small stem and leaf-petiole cuttings.

Use any type of glass or transparent container you'd like to water root your cuttings or seeds. Glass jars, vases, test tubes—a martini glass—you name it. They all work. There are also a few water rooting accessories you can use for more successful water-propagation.

Ceramic and porcelain sprouting supports sit on top of a glass or jar, providing support for a large seed or vegetative cutting. Providing this support means the seed or cutting won't sink into the water where it could drown and rot. These sprouting supports are also handy for rooting and forcing bulbs.

The old-school method for germinating avocado seeds in water is to stick three toothpicks into the seed, then suspend it on top of a glass jar. If you're looking for a more modern approach, look for avocado seed floaters, which hold the seed and float it in a bowl of water.

When water rooting, be sure to change out fresh water as it becomes cloudy, until the plant has developed adequate roots. The type of water you use can make a difference in how well your plant roots and grows. City tap water will contain chorine, which is toxic to plants. Burnt looking or brown edges along

Assorted stem and leaf cuttings growing roots directly in water with ceramic sprouting supports.

∧ Did you know you can root stems of succulents and cactus in water? Just make sure the fleshy part of the plant is kept above the water level. You can accomplish this using a container with a narrow neck, ceramic rooting dishes, cutting holes in plastic lids, or toothpicks—it's a DIY adventure.

∧ These cute avocado-shaped boats float your avocado seeds in water.

your plant leaves can be a sign of chlorine toxicity. Tap water will also often have softeners such as sodium, which will damage your plants. If you can collect rainwater, you'll find it to be ideal for rooting new cuttings and seeds. You can use spring water or well water, too. That said, if you only have access to tap water, go ahead and use it. You can naturally dechlorinate tap water by setting it out in a clear container in natural light for at least 24 hours.

The time it takes to root in water simply depends on the type of plant you're growing. Certain fast-rooting species can root in only a few days; most plants will take a couple of weeks to develop roots, while others can take more than a month. Fleshy plants that develop aerial roots, such as philodendron are particularly easy and fast to water root. Plants with tougher stems will take longer. You can add small concentrations of rooting hormone, or root stimulators, to water that you use to root new plants. Be sure to follow label instructions for how much to use when mixing with water.

Once plants have developed roots that are 1 to 2 inches long and begin branching, it's time to pot up your plant into a potting mix.

This tip cutting of a chartreuse heartleaf philodendron rooted in less than a week.

Taking herb tip cuttings.

VEGETATIVE

Propagation

How to Clone Your Plants Without Seeds

This whole-leaf cutting of an echeveria will grow into a clone of its mother plant. >

How do you propagate a plant when you don't have seeds or when you want to speed up the propagation timeline? Vegetative cloning. It can take anywhere from 8 to 12 weeks to grow a garden-ready or big container–ready tomato transplant from seed, but you can root and start a vegetative cutting (clone) from an existing tomato plant in a quarter of that time. Each type of plant will have its own time frame. Tomatoes, for example, can start producing new roots from vegetative cuttings in a matter of days, while citrus or woody shrubs can take much longer—even months.

Another good reason for choosing vegetative propagation is when you want exact copies of the same plant you already have. You can clone your plant by taking vegetative cuttings, a piece of a stem, or other part of the plant and rooting it. Vegetative propagation requires access to a viable mother plant from which to take cuttings, as well as learning a few basic techniques. You might own a plant that was special in your family or handed down from a relative. Cloning such plants is a wonderful way to keep their legacy alive as well as gift exact copies of your special plant to family and friends.

WHERE TO START

Different plants have varying requirements for how, where, and when to take cuttings, the ideal size of the cuttings, and how long they may take to root. Some plants, such as citrus or ivy, will root from stem and leaf-bud cuttings. You can propagate many tropicals, herbs, and perennials, such as pothos ivy and salvia, by placing a plant stem cutting in water for a few weeks. Many succulents will root from stem cuttings, but also can root from a single leaf (whole-leaf cutting). Some plants, such as peperomia, begonia, and African violets, will root from a leaf with a section of petiole attached (leaf-petiole cutting), and can also root straight from the leaf veins (non-petiole/split-vein cutting).

Some plants have evolved to produce small plant-lets, or offsets (pups), that grow and develop their own root system while still attached to the mother plant. Many perennials are propagated by taking root cuttings

∧ These African violet leaves are growing new roots from a leaf-petiole cutting.

∧ This versatile peperomia plant can be propagated several ways, including stem-tip cuttings, leaf-petiole cuttings, split-vein cuttings, and even by offsets and division. The mother plant is healthy and a good source of cuttings.

This tiny echeveria offset, or pup, was growing at the base of its mother plant. >

or divisions. Some plants, such as citrus, will even produce asexual seeds that are clones of the parent plant. These clone seeds are essentially a type of vegetative reproduction. Clearly, you must get to know a few things about the plant you want to clone before you get started.

When taking cuttings, be sure to start with a vigorous, healthy mother plant that has characteristics you want to replicate. The overall health and nutrient levels of the mother plant will have a big impact on the success of your cuttings. Only take cuttings from plants that are insect- and disease-free, and that don't show any nutrient or moisture stress. Also avoid taking cuttings of tissue with flowers. New cuttings need to establish a good root system before pushing out new leafy growth that they are not mature enough yet to support. Cuttings tend to root better and take off faster if the mother plant has high levels of carbohydrates and less nitrogen—stop fertilizing your mother plant with nitrogen for about a week before you take cuttings. Cuttings from younger plants tend to be more successful than cuttings taken from older more mature plants. Sometimes cuttings from lateral shoots will root better than those taken from terminal shoots.

Something else to consider is that different plants root better depending on the age of the growth.

SOFTWOOD CUTTINGS Cuttings taken from the soft, new tip growth of annuals, herbs, perennials and some woody plants, such as pothos ivy, basil, verbena, pelargonium, and hydrangeas, and many tropical houseplants, before it matures and hardens.

SEMI-HARDWOOD CUTTINGS Cuttings taken from semi-mature sections of the current season's growth on plants such as rosemary, citrus, English ivy, and camellias; the tissue is still flexible but beginning to stiffen.

HARDWOOD CUTTINGS Cuttings taken from dormant sections of mature stems of shrubs and trees, such as hollies and junipers, in the late fall or winter. There is no active growth on these cuttings and they are harder stems.

The time of year you take cuttings from certain plants can influence their success. It's usually better to take stem cuttings from plants when they are most actively growing. For many plants in your outside garden, the prime time to take cuttings is spring, early summer, or early fall. Early morning is the best time of day as your plant is usually fully turgid (full of water).

The key is to be efficient and take advantage of what the plant has already grown and developed. For example, when you take a stem or leaf-bud cutting, the tissue will only need to develop new adventitious roots, as the cutting already has pre-existing shoot tissue. Offsets and runners already have adventitious root systems, and so can be very quick and easy to propagate. When you divide plants at the root zone, you're essentially separating a new complete plant—it will just need a little time to grow some new root tissue. A root cutting or a leaf cutting, on the other hand, must initiate both new adventitious roots *and* a new adventitious shoot; that means they will take longer to form new plants than stem and leaf-bud cuttings, offsets, and divisions.

Some species of plants are not easy to clone vegetatively. Certain pome fruits, such as peach, apple, pear, and cherry, as well as other woody plants, don't root well from cuttings. They are typically propagated by grafting and budding, which are more advanced propagation techniques not covered within the scope of this book. Such plants can be challenging for beginner propagators. Plants such as irises cannot be propagated from leaf cuttings, but you can easily divide their rhizomes. In this chapter, we'll cover the basic methods of vegetative propagation to get you started with a variety of common plants.

You will find there is some overlap between the following vegetative propagation categories, because some plants have multiple ways they can be propagated. Even so, I've broken them down into a few simple categories to help you learn plant cloning step-by-step.

STEM CUTTINGS

When it comes to plant cloning, you'll probably take more stem cuttings than any other type. There are a few different methods you can use to take stem cuttings depending on your plant of choice. Many small houseplants are easily propagated by stem-tip cuttings, while large tropical specimens can be air layered. Here we'll cover the different types of stem cutting methods and corresponding plant types.

STEM-TIP CUTTINGS

Once you get down to making more plants, you'll do most of your cloning using stem-tip cuttings—also called tip cuttings—especially if you love water rooting. Tip cuttings are taken from plants with distinct stem tissue that can be cut away in sections from the mother plant. Plants commonly propagated via this method include basil, pothos ivy, mint, philodendron, salvia, geraniums, coleus, and many succulents.

⌃ Many tropical houseplants root easily from stem-tip cuttings.

Tip cuttings of oregano, watercress, lavender, thyme, rosemary, and a different variety of oregano. I use small sharp snips to remove a 2-inch section that contains two or three nodes along the stem. Each of these herbs will typically root directly in water, if I choose not to root them in a growing media. ❯

You can also stick tip cuttings in a rockwool propagation tray.

⌃ Tip cuttings can be rooted in a lightweight growing media. Dip cuttings in rooting hormone before sticking.

⌃ A tip cutting of a corn plant (*Dracaena* spp.) stuck directly into a 4-inch pot with moist growing media. In a few weeks, this cutting will develop new roots and then begin to grow new leaves from the growing tip.

⌃ After a few weeks in an aeroponic propagator, this citrus cutting has developed new crown roots. As soon as these roots begin to branch (develop small lateral roots) it will be ready to plant.

This stem-tip cutting of a fiddle leaf fig can be rooted in water, moist growing media, rockwool, or Oasis plugs. Because of the size of its leaves, it's best to remove all the lower leaves—or all the leaves—before you stick the cutting. ⌄

How to Take a Stem-Tip Cutting

1   Disinfect your snips using alcohol or a 10 percent bleach solution.

2   Make a clean cut about 4 inches below a relatively new stem tip that has no flowers or buds on it. Take a 3- to 5-inch cutting from a plant stem, then remove any additional side stems and the bottom leaves. After removing leaves, you can trim the cutting to 3 to 3 ½ inches, snipping just below a leaf node. Be sure that the remaining cutting includes at least one leaf node, which is where the new shoot will emerge from the main stem. Be careful not to damage the stem base, as this is where the callus and adventitious roots will form.

3   Dip the bare end of the stem into the rooting hormone.

4   Gently place or stick the cutting into water, rooting growing media, rooting plug, or an automatic propagator system.

5   Tropical plants and plants that prefer more moisture and humidity may root faster if covered until rooted. If you are rooting cuttings in organic or inert growing media—or a propagator—cover your cuttings with a humidity dome or other plastic cover. (If you are water rooting, you don't need to cover the cuttings.) If your humidity dome has a vent, you can crack it open if water is condensing on the plastic. Don't overcrowd—while you must maintain humidity around the cuttings during rooting, they also need good air circulation. Once the cuttings have started developing roots, you can remove the dome or cover.

6   Keep the rooting media moist, but not soggy.

7   Place your cuttings near natural light from a window or a couple of feet away from fluorescent, CFL, or LED grow light fixtures. Intense light will burn cuttings. If using grow lighting, leave lights on for 24 hours a day.

8   Give it time. Some plants can take a couple of months to root, while others root in a matter of days. For plants that like warm growing conditions, or if you're starting cuttings in winter or in a cool part of your home, use a seedling heat mat under the rooting tray to speed things up.

9   Once roots have filled the original container, it's time to bump up your plant to a slightly larger container.

∧ These succulent stem-tip cuttings, or muffin-tops as I like to call them, were set out for more than a month to cure.

Because succulent plants contain so much moisture, it's best to let cuttings cure for at least a few days before inserting into water or growing media, so the wounded area of the cut can begin to seal and callus. Once your succulent tip cuttings have cured and begun to callus, you can set the stem in water or into a slightly moist growing media or plug. Don't be concerned if you notice a bit of drying or shriveling on some of the leaves—succulents are resilient. In fact, you can often leave succulent cuttings out for extended periods of time until they develop new root tissue, then put them in water or pot them up. Succulent cuttings do not need to be covered.

When taking stem-tip cuttings, you will find that different plants will require you to remove more or less foliage from the new cutting. Tomatoes, salvia, mint, and pothos ivy, for example, are fast, vigorous stem rooters. You can get away with leaving more foliage on the cutting.

Slow-to-root and woody cuttings perform better if you remove more leaves. When you leave excess foliage on slow-to-root cuttings, the chances of fungal problems with surrounding cuttings can increase. Excess leaves can yellow and drop off the cutting. The excess leaves also cause the cutting to lose more moisture and take up resources. So, I usually remove all but one or two of the top leaves from slower cuttings, such as peppers and citrus, before sticking them. For many woody types of plants, such as roses, you may remove all the leaves from the cutting.

⌃ An echeveria tip cutting has calloused and developed new roots. At this stage, it's ready to be potted up into a small container with growing media.

⟨ Rose cuttings will take longer to root than tropicals or succulents, so they will benefit from increased humidity as well as a heat mat. I typically remove most or all of the leaves from rose cuttings, as rose leaves can be susceptible to fungal diseases in higher humidity environments.

⟨ Normally, you'd remove more leaves from your cuttings—but these tomato cuttings root quickly, so you can leave more foliage intact.

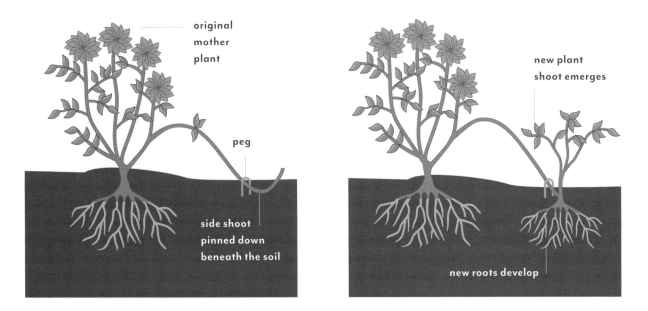

original mother plant

peg

side shoot pinned down beneath the soil

new plant shoot emerges

new roots develop

∧ You can layer plants in your garden by pinning down side shoots to the soil.

GROUND AND AIR LAYERING

Ground layering, also called simple layering, is a straightforward way to propagate plants that easily grow new roots along their stems. Layering occurs in nature when a branch bends down or falls over and touches the ground. If it stays in contact with the soil, adventitious roots can develop and anchor into the ground. Eventually the newly rooted shoot is separated from the parent plant (or the parent plant may die) and the rooted shoot becomes a new plant specimen.

You can replicate this process in certain plants by bending a branch down and pinning it to the ground so a piece of the stem meets with ground soil or into another pot of soil. You can also bury the tip of the branch (called tip layering), so it will first grow downward and then back up out of the soil.

Depending on the plant, new roots can take anywhere from several weeks to a year to develop. You can speed up the process by making a small wound on the branch to expose the inner tissue and applying some rooting hormone to the area before pinning it to the ground or soil. After the shoot has rooted, you cut it away from the parent plant and it will continue to grow on its own.

Brambling plants, such as blackberries and raspberries, are commonly layered, but this method also works for many other plants, including fruit trees, tropicals, vines, azaleas, forsythia, rhododendron, and roses.

You can grow dragon fruit (*Hylocereus* spp.) from seed, but germination is unpredictable and can take a long time. Vegetative cuttings, on the other hand, are easy. This young, fleshy stem of a succulent dragon fruit bent over and touched the soil in the pot. The stem, where it touched the soil, produced new adventitious roots and a new bud shoot. I can now cut this shoot away from the stem on either side and transplant it. You can also cut the fleshy stem into several sections, let them cure, then pot them up individually to root. >

If you are focused on propagating houseplants and tropicals in containers, then you can try air layering. **Air layering**, or air propagation, is similar to rooting with ground layering or a stem-tip cutting, except you don't bury the stem in soil or remove a cutting from the mother plant; you root it right on the plant while it's growing! You accomplish this fascinating feat by creating an artificial root zone right along an actively growing stem. If the cells along the stem think they are under soil or in water, they will start producing new roots—just as with ground layering.

Citrus trees and large tropical houseplants are good candidates for air layering. You can also air layer shrubs and trees (fruit trees are a popular choice) in your outdoor garden. The best time to start an air layer cutting is in early spring when the sap inside the plant begins flowing—especially with fruit trees. If you plan to air layer deciduous plants outdoors, watch for the first sign of their buds turning green. Indoor tropical plants can be air layered any time of year they are actively growing.

You can air layer a plant on its side branches or even on its main trunk. This is a great way to rejuvenate ragged houseplants that have become overstretched or overgrown. Instead of throwing away the plant, you can air layer it along the main trunk, preserving the healthy top growth. Once a new root mat has developed, simply remove and pot up the cutting—like new!

My fiddle leaf fig has gotten a bit lanky. So, I've air layered it on a side branch. Once roots develop to the edges of the air-layering ball, I'll cut off the entire section and pot it up as a new plant. >

⌃ Air layering is particularly easy when using a kit that comes complete with a ball that snaps together. You can also use clear plastic wrap to hold sphagnum moss or coir in a ball around your air layer cut. I used coir in the layering ball shown in the photos.

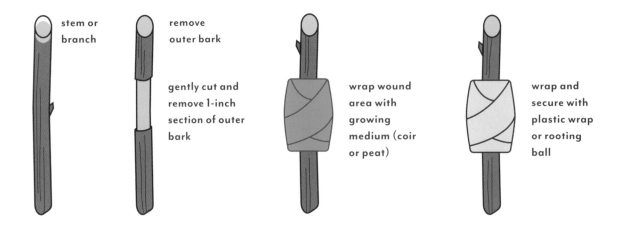

stem or branch

remove outer bark

gently cut and remove 1-inch section of outer bark

wrap wound area with growing medium (coir or peat)

wrap and secure with plastic wrap or rooting ball

∧ Step-by-step process of air layering.

How to Take an Air Layer Cutting

1 Disinfect your knife using alcohol or a 10 percent bleach solution.

2 Choose a branch (either side stem or main stem) that is at least 3/8 inch thick to girdle. Cut through the external bark layer, all around the circumference of the stem. Then make a second identical cut on the same stem. The distance of the second cut should be about one and a half times the diameter of the branch/stem.

3 Make a couple of perpendicular cuts between the two original cuts so that you can peel the bark away from the section. Essentially, you're removing a section of "skin" to stimulate the plant to generate new tissue. On plants that don't have thick bark, you can simply scrape away the outer tissue between the cuts.

4 Use the edge of the knife to scrape away some of the cambium layer. This stops the flow of water and nutrients from the main part of the plant into the stem section. This is where your new roots will grow.

5 Soak some sphagnum moss or coir in a bowl of water overnight. Squeeze out extra moisture before you use it.

6 Cover and protect the cut area by packing a mat of moist sphagnum moss or coir around the

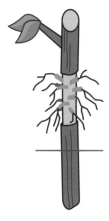

when roots are visible and fill growing media, remove plastic or ball; leave growing media attached to roots

cut just below rooted area

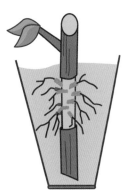

transplant rooted cutting into a container or into the garden

girdled area. The moss mat should be about the size of a baseball. Make sure it completely covers the girdled section.

7   Use clear plastic wrap around the moss mat, wrapping tightly, so that the moss is in firm contact with the stem and there are no loose areas. Seal both ends, like a candy wrapper, with twist ties so that all the moisture is contained. You can also use an air layering kit with a ball that snaps together.

8   After the girdled section is wrapped up tight, it's best to trim off any small side branches growing above the air layered section but leave any side branches that are growing on the top

6 inches of the stem. The section of stem inside the plastic wrap or layering ball does not need any light exposure.

9   Once you see new root tips growing all the way around the moss, it's time to cut your new plant away from the mother plant. This process can often take 2 to 3 months—longer for certain types of plants. To harvest your new cutting, cut the branch below where the new root ball has formed and pot it up.

∧ The step-by-step process to taking a cane cutting from a corn plant (*Dracaena* spp.).

CANE CUTTINGS

Cane cuttings are an easy way to propagate many tropicals that grow thick, sectioned stems, such as Chinese evergreen (*Aglaonema* spp.) or corn plant (*Dracaena* spp.). You can also propagate certain types of orchids that grow upright (monopodal), such as *Vanda* spp. and moth orchid (*Phalaenopsis* spp.), by removing a section of stem/cane.

If you have houseplants that have become leggy and stretched in low light, it may be time to start new ones. With cane cuttings, you'll remove a 2- to 3-inch section of older main cane stems without any leaves. You can root cane cuttings either horizontally (if you don't want a cut section of cane exposed on your new plant) or vertically. Once the cane cutting develops roots and a new bud shoot, you can let it fill out the container and then repot it into a larger container.

∧ This cane cutting has rooted and is growing new shoots.

How to Take a Cane Cutting

1 Disinfect your snips using alcohol or a 10 percent bleach solution.

2 Choose an older section of a leafless main stem that includes two or three nodes and cut it from the plant. Make flat cuts—not at an angle. Cut away top growth with any leaves.

3 Dip the bottom end (vertical cutting) or the bottom half (horizontal cutting) into the rooting hormone.

4 If rooting horizontally, simply lay the cutting on top of the rooting media and press down a bit so it makes contact with the media. If rooting vertically, stick the cutting into rooting media or plug, ensuring that about half of the cutting is beneath the surface. You can also vertically root cane cuttings in water or propagators.

5 Cane cuttings typically benefit from use of a small humidity dome or a clear plastic bag, although some cane cuttings will root and shoot perfectly fine uncovered. Do keep the growing media moist if you're using potting mix or substrates such as Oasis or rockwool.

6 Then follow steps 7 to 9 in "How to Take a Stem-Tip Cutting" (page 144).

STEM LEAF-BUD CUTTINGS

Stem leaf-bud cuttings are similar to stem-tip cuttings, except you make two cuts to remove sections from along the plant stem—usually from plants with long trailing stems, such as ivy, philodendron, jasmine, and clematis. Tropicals such as rubber and jade plants, herbs such as basil and mint, and many shrubs such as camellia, blackberry, and citrus are also propagated using this technique.

Stem leaf-bud cuttings include a leaf, petiole, and a small section of attached stem. You take a leaf-bud cutting by cutting both below and above a leaf node. The piece of stem that is below the leaf node (closest to growing crown of the plant) is what you will insert into water or growing media. Or, you can place it horizontally into the soil, with the leaf and node above the soil (like a cane cutting). New roots will develop from the cut portion of the stem and a new shoot will emerge from the leaf node.

∧ Stem leaf-bud cuttings taken from a rose plant. All the leaves have been removed.

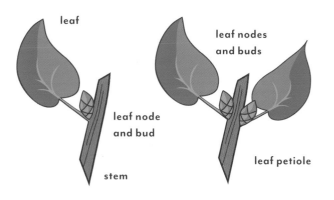

leaf

leaf node
and bud

stem

leaf nodes
and buds

leaf petiole

﹤ A stem leaf-bud cutting is a section of stem cut on both ends, containing one or two leaf nodes. One leaf-bud cuttings include one leaf blade, leaf petiole, and a short piece of stem. There is a leaf node (joint) between the leaf petiole and the stem. Two leaf-bud cuttings have two nodes, one on either side of the stem, between the leaf petioles and stem.

A stem-tip cutting of a fiddle leaf fig on the left taken from the top of the stem, and a stem leaf-bud cutting on the right. The leaf-bud cutting is a section of stem cut on both ends, containing one or two leaf nodes.

How to Take a Stem Leaf-Bud Cutting

1   Disinfect your snips using alcohol or a 10 percent bleach solution.

2   Make a clean cut about 4 inches below a relatively new stem tip that has no flowers or buds on it. Then take a 2- to 3-inch cutting from along the plant stem, making the first cut at an angle about ¼ inch above the leaf node and then a second straight cut 1 to 2 inches below the leaf node. Be sure that the remaining cutting includes at least one leaf node, which is where the new shoot will emerge from the main stem. Be careful not to damage the stem base as this is where the callus and adventitious roots will form.

3   Remove all but one or two leaves.

4   Gently place or stick the cutting into water, rooting growing media, rooting plug, or an automatic propagator system.

5   Leaf-bud cuttings can benefit from using a humidity dome, plastic cover, or plastic bag to increase humidity. Keep the rooting media moist, but not soggy.

6   Then follow steps 7 to 9 in "How to Take a Stem-Tip Cutting" (page 144).

LEAF CUTTINGS

Taking leaf cuttings is an easy and fun way to propagate new plants. No matter how many times I do it, seeing a tiny new plant growing right from the base of an old leaf is always exciting. You'll take leaf cuttings from plants that are very leafy with only small sections of short stems, or fleshy succulents with the ability to form new roots and buds directly from leaf and leaf petiole tissue. We'll discuss how to take several different types of leaf cuttings.

WHOLE-LEAF CUTTINGS

For whole-leaf cuttings, look for plants that lack petioles with the leaf attaching directly to the main stem. Succulents are a great example of a type of plant from which you can easily take whole-leaf cuttings—just be sure to allow leaves to develop a callus.

∧ Succulents often have leaves that you can detach from the main stem and root as whole-leaf cuttings.

These whole-leaf cuttings of assorted succulents are resting on dry potting mix, where they will dry and develop a callus at the base of the leaf before developing new adventitious roots and bud shoots. >

∧ Whole-leaf cuttings of echeveria plants, first forming a callus, then new root initials, then new bud shoots, which will form the new plant.

∧ Echeveria leaves in various stages. From callusing, to root development, to new bud shoots and a developing pup.

How to Take a Whole-Leaf Cutting

1 You'll take most whole-leaf cuttings from succulent plants. Gently snap off a leaf at the base where it attaches to the main stem. Take care not to break away any of the leaf tissue at the base or it may not be able to form a callus or new root tissue.

2 Set succulent leaves out on a dry, hard surface to cure. You can also rest succulent leaves on the surface of a totally dry growing media to cure. The area attached to the stem will begin to seal and form a small callus.

3 If you've cured the callused succulent leaves on a hard surface, now place them directly on top of a loose succulent potting mix or a 50/50 mixture of builder's sand and coir in a flat or other shallow container, where they will begin to develop new adventitious roots. You also can set leaf cuttings on top of mix in a small pot with a drainage hole.

4 It's best to not cover succulent leaf cuttings. You can lightly mist the soil and cuttings after they begin to develop new roots. Once small bud shoots and plantlets begin to appear, you can water lightly using a water dropper or squeeze bottle.

∧ These whole-leaf cuttings of an echeveria are growing new bud shoots. The parent leaf, the original leaf cutting, is beginning to dry and will eventually fall away from the new plantlet.

5 Place your cuttings near natural light from a window or a couple of feet away from fluorescent, CFL, or LED grow light fixtures. Intense light will burn cuttings. If using grow lighting, leave lights on for 24 hours a day.

6 Eventually, as the new plantlet continues to grow, the original leaf cutting, also called the parent leaf, will dry up and fall away. At this point, you can transplant your new succulent from the propagation tray into a small container or allow it to continue to grow in its small pot.

∧ Many favorite houseplants can be propagated with leaf-petiole cuttings.

LEAF-PETIOLE CUTTINGS

Not all plants can develop adventitious roots and buds directly from the base of a leaf—many need a section of petiole attached or a piece of the main stem. Plants that can generate new roots and bud shoots from the leaf petiole (the stalk that connects the leaf to a main plant stem) can be propagated from leaf-petiole cuttings, also called leaf-stem cuttings. Leaf-petiole cuttings are better for plants that are very leafy with only small sections of short stems, such as begonias, peperomias, and African violets, as well as other gesneriads. You can root leaf-petiole cuttings in potting mix, water, propagators, or in other inert substrates such as Oasis.

< ∧ I rooted this begonia leaf-petiole cutting in water on my windowsill. After it rooted and developed a baby plantlet, I cut away the plantlet and potted it up. You could also pot up the leaf petiole after it develops roots and then the baby plantlet will emerge from the soil.

How to Take a Leaf-Petiole Cutting

1 Disinfect your snips using alcohol or a 10 percent bleach solution.

2 Choose a healthy leaf and make a clean cut on the petiole, leaving about 1 inch of the petiole attached to the leaf. Cut the petiole at a 45-degree angle with the cut side facing up to encourage more root and bud production.

3 Optional step: You can use a sharp razor blade to cut off the top half of the leaf blade. This can encourage faster rooting—but is not necessary.

4 Dip the bare end of the petiole into rooting hormone.

5. Gently place or stick the petiole into water, rooting growing media, rooting plug, or an automatic propagator system. Make sure the petiole is in firm contact with the surrounding growing media.

6 Many plants from which you typically take leaf-petiole cuttings may not require a humidity dome or plastic cover, as many are susceptible to fungal diseases. Do keep the growing media moist, especially if you're using potting mix or substrates such as Oasis or rockwool. Use heat mats to speed up rooting.

< These African violet leaves were rooted in water. You can see both the new adventitious roots and a tiny bud shoot developing at the base of the petiole. You can pot up these rooted petioles, placing the roots below the soil. The new shoots will later emerge from the soil.

7 Place your cuttings near natural light from a window or a couple of feet away from fluorescent, CFL, or LED grow light fixtures. Intense light will burn cuttings. If using grow lighting, leave lights on for 24 hours a day.

8 New roots and bud shoots will develop at the base of the petiole and emerge through the soil. If you are water rooting leaf-petiole cuttings, you'll need to pot up the cutting after a small root mat develops at the base of the petiole.

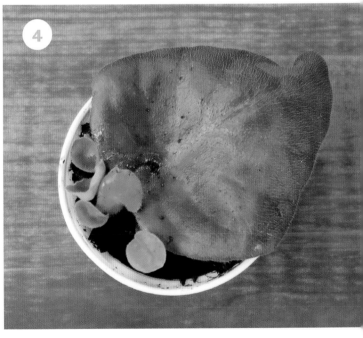

An African violet leaf-petiole cutting rooted in moist growing media. You can see small bud shoots developing from the base of the original leaf, which will decay and fall away.

NON-PETIOLE AND SPLIT-VEIN CUTTINGS

Not all plants need a stem or leaf petiole to sprout new plants. Some plants can develop both roots and bud shoots with only a piece of a leaf (non-petiole cutting) or directly from leaf-vein margins along the leaf (split-vein cutting). When the leaves of plants with this ability become damaged or a piece of leaf falls to the ground, they might start developing new roots and shoots from the leaf, unless the leaf rots first. There are a few different methods for taking non-petiole and split-vein leaf cuttings, and different types of plants respond differently to each method.

Plants that respond well to split-vein cuttings include plants with large fleshy leaf veins, such as African violets, rex begonias, pilea, *Gloxinia* spp., and *Smithiantha* spp.

◁ I pressed this begonia leaf-petiole cutting into a small pot of moist growing media. After a couple of months, a new leaf-bud has emerged from beneath the soil.

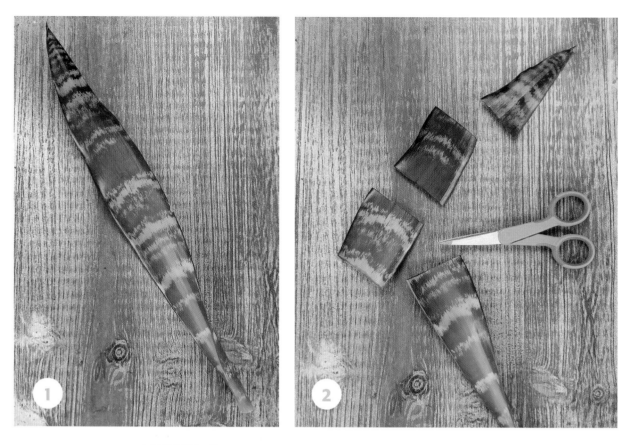

You can cut snake plant (*Sansevieria* spp.) leaves into multiple sections; new roots and shoots will develop from the base of each leaf cutting.

◁ African violet non-petiole leaf cuttings rooting under a humidity dome.

①

②

③

How to Take a Non-Petiole Leaf-Section Cutting

1 Disinfect your snips using alcohol or a 10 percent bleach solution.

2 Choose a healthy leaf from your mother plant.

3 Cut off a small 1- to 2-inch section of the leaf with a horizontal cut (plants that respond well to this method include African violet, begonia, pilea, snake plant, and eucomis); you can cut larger leaves into multiple sections. Or, cut the leaf in half along the midrib, then discard the midrib so you're left with two leaf sections (plants that respond well to this method include African violet and *Streptocarpus* spp.).

4 Dip the cut end into rooting hormone. For leaves cut into multiple sections, dip the bottom cut end (originally closest to the base of the plant) into rooting hormone.

5 For leaves cut horizontally, gently place or stick the leaf section into water or the rooting media. Make sure the cutting is placed in the same direction in which it was growing on the plant. For leaves cut along the midrib, create a small trench in the growing media and gently insert the cut side of the leaf cutting. Make sure that both types of cuttings are in firm contact with the growing media.

< If you look very closely, you'll see a tiny plantlet shoot developing on top of the leaf where it was damaged.

6 Non-petiole cuttings from fleshy or succulent plants typically root just fine uncovered, while more tropical plant types can benefit from a humidity dome, plastic cover, or plastic bag. Keep the media moist but not soggy. Remove the cover if it appears the cutting is too wet or beginning to decay. Use heat mats to speed up rooting.

7 Place your cuttings near natural light from a window or a couple of feet away from fluorescent, CFL, or LED grow light fixtures. Intense light will burn cuttings. If using grow lighting, leave lights on for 24 hours a day.

8 New bud shoots, or plantlets, will develop along the cut section of the leaf. Eventually, the original parent leaf section will die and fall away. You can cut away or lift out the new plantlets for repotting.

How to Take a Split-Vein Cutting

1 Disinfect your snips using alcohol or a 10 percent bleach solution.

2 Choose a healthy leaf from your mother plant.

3 Make small cuts through some of the main veins on the underside of the leaf. You may also cut off the leaf margins (the edges of a leaf) if you'd like. Dab rooting hormone on vein cuts.

4 Place the leaf flat on the surface of a loose potting soil/growing media, making sure that the top of the leaf faces up and the cut veins underneath make contact with the growing media. You may need to use toothpicks or other small weighted objects (glass beads, small pebbles) to hold the leaf in place.

5 Split-vein cuttings can benefit from use of a humidity dome, plastic cover, or plastic bag to increase humidity. Don't let media or the leaf stay soggy when under plastic. Remove the cover if the cutting appears to be too wet or is beginning to decay. Use heat mats to speed up rooting.

6 Place your cuttings near natural light from a window or a couple of feet away from fluorescent,

∧ A split-vein begonia cutting sits flat on the soil surface.

CFL, or LED grow light fixtures. Intense light will burn cuttings. If using grow lighting, leave lights on for 24 hours a day.

7 The new plantlets will develop roots and shoots where the veins were cut. You can cut away, or lift out, the new plantlets for repotting.

∧ The petiole of this fiddle leaf fig plant can develop roots, but it won't develop a new bud shoot.

∧ Leaf-petiole cuttings of Chinese money plant (*Pilea peperomioides*) don't always develop new buds, only roots. It's best to use tip cuttings, pup cuttings, or division.

BLIND CUTTINGS

One important thing to know about leaf cuttings is that just because a certain plant can easily develop a callus or adventitious roots on a leaf, petiole, or stem section, it may not be able to develop a new adventitious bud or shoot from that same location. In fact, some leaf cuttings can survive on their roots for a very long time, yet never develop a new bud or shoot. We call these blind cuttings. Some parts of the plant can generate some forms of new tissue, but others cannot.

This is true for certain species of ficus, such as the classic rubber plant (*Ficus elastica*) and the oh-so-trendy fiddle leaf fig (*F. lyrata*), as well as certain species of hoya (*Hoya kerrii*) and jade (*Crassula* spp.). Sure, you can pop off a whole leaf or a leaf with a piece of petiole attached and it will develop roots in water or soil—but that's all it will ever do. For these plants, a leaf-petiole cutting won't be enough, as the leaf-petiole cutting will almost never form a new bud on its own. You must take leaf-bud cuttings, a piece of stem with a leaf node included, for these plants to grow a new whole plant.

I've also experienced this challenge with leaf-petiole cuttings of Chinese money plant (*Pilea peperomioides*). The petiole developed a callus and roots, but the cuttings never developed new buds or plants. So frustrating! Certain chemicals can be applied to such blind cuttings to help them develop a new bud shoot, but for the home gardener, it's best to pitch blind cuttings into the compost bin and try a new cutting method.

OFFSETS

There are some very considerate plants that do most of the propagation work for you in advance. Such plants produce small, fully functional daughter plants complete with root systems at the end of a long flowering stem, at the base of the crown, along the stem at leaf axils, or even along leaf margins. These tiny clones go by a few names, such as offsets, pups, plantlets, or runner plantlets. This is an effective way for a plant to clone itself, especially if a compatible pollination partner doesn't happen to be nearby. Potting up these pups is also a really easy way to make more of the plants you love.

Airplane plant is a classic tropical houseplant that produces runner plantlets en masse, making it a favorite for home propagating. You'll notice these plants develop long flower shoots and, at the ends of these shoots, you'll find several runner plantlets. Moth orchids occasionally produce baby orchid plantlets—specifically called keikis when referring to orchids—off their main stem, or sometimes on the flowering stem. Often these plantlets will continue growing larger on the stem. As soon as a plantlet comes in contact with water or soil, it will develop new root tissue and continue growing. All you have to do is snip it off and either set in water to stimulate root growth or pot it up in a small container with potting mix.

∧ Airplane plant is one of the easiest houseplants to clone. It produces runner plantlets at the end of flowering stems, complete with undeveloped roots. Simply place in water to stimulate root growth, then pot up once the roots are 2 to 3 inches long.

Strawberries are another classic runner plant, sending out long runner shoots (stolons) that contain small plantlets with tiny node roots just waiting to touch the soil.

How to Take a Runner Offset Cutting

1 Disinfect your snips using alcohol or a 10 percent bleach solution.

2 Find a runner plantlet along a flowering or main stem and cut it away from the mother plant. Make the cut close to the plantlet but don't cut into its growing crown or roots.

3 Dip the base into rooting hormone.

4 Set the base of the runner plantlet in water to stimulate new root growth (in which case, leave in water until roots are 2 to 3 inches long) or set it in a plug tray.

5 Runner plantlet cuttings typically benefit from use of a humidity dome or plastic cover for a short time; however, many will root and shoot just fine uncovered. Do keep the growing media moist if using potting mix or substrates such as Oasis or rockwool.

6 Then follow steps 7 to 9 in "How to Take a Stem-Tip Cutting" (page 144).

Many succulent plants, including aloe, agave, bromeliad, haworthia, and pilea, develop small pups on underground stems that are close to the soil surface—or even directly on the stem. You'll see them popping up in the soil at the base of the mother plant. Some air plants, such as bromeliads, develop pups right on the crown above the root zone. You can gently pull or cut these pups away from the mother plant (retaining some root initials).

It's best to let newly cut succulent plant pups dry and callus for a few days before water rooting them or placing them in plug trays or small containers with growing media. If you're removing pups from plants out in the garden, and indoor potted plants, the best time to do so is late spring and early summer, when plants are most actively growing.

Many orchids grow upright (monopodal) and produce small keikis (offsets) along their stems that you can remove and pot up in the same way. You can force the growth of keikis by topping an upright orchid that has grown tall and has started producing lateral aerial roots along the stem—much like taking a stem-tip cutting. Simply cut the stem tip below where the roots are growing and pot it up. The orchid will then begin to develop keikis along the lower part of the stem that remains on the original mother plant.

Some sideways-growing (sympodial) orchids, such as lady slipper orchids (*Paphiopedilum* spp.), produce offsets around the main growing stem at the soil line. These offsets can be divided the same way as succulent pups.

< Most types of
strawberries produce
runners with offsets
that include node
roots.

< This haworthia
succulent has devel-
oped several small
pups around the base
of the crown that
can be removed and
planted. I removed
the haworthia mother
plant from the soil so
you could clearly see
the small developing
pups, which are a
lighter green color.

Chinese money plant (*Pilea peperomioides*) is most easily propagated by removing offsets (pups) that grow around the central stem of the plant as well as on it. Expose one of the pups so you find roots growing from the base of the stem and cut it away from the mother plant using sharp snips. Then pot up the pup into a small container with moist growing media. In a couple of weeks, the pup will start growing and producing new leaves as it roots.

This lady slipper orchid has grown a new offset along its main stem that can be cut away and potted up. Normally, I wait until the plant has finished flowering before I take the offset, but I want you to see the main stem with the flower.

4

How to Take a Pup Offset Cutting

1   Disinfect your snips, sharp knife, or razor blade using alcohol or a 10 percent bleach solution.

2   Find a pup growing around the base of the plant crown. Pull away the soil from around the pup if you've removed it from beneath the growing crown. If you are repotting houseplants, you may take the plant completely out of the pot and brush away soil to expose pups. Cut away the pup, making sure to include part of the underground stem that has some root initials. If working with an air plant, you can usually gently pull the pup away from the crown with your hands.

3   Allow pups of succulent plants to callus for several days.

4   Set the base of pups in water to stimulate new root growth (in which case, leave in water until roots are 2 to 3 inches long) or set in small pots or plug trays filled with a loose growing media.

5   Most pup cuttings will root just fine uncovered but do keep the growing media moist to the touch.

6   Then follow steps 7 to 9 in "How to Take a Stem-Tip Cutting" (page 144).

PSEUDOBULBS AND BULBILS

Many popular orchids aren't propagated from seed or cuttings, but rather from pseudobulbs. A **pseudobulb** is a pod-like structure that grows right below a leaf—usually along a horizontal growing stem that may be under growing media. You'll find pseudobulbs on sympodial-type orchids. The small pod contains nutrients and water, just like an underground bulb, acting as a resource backup for the main plant when it's under stress—but each pseudobulb has the potential to become a new plant. When the main plant taps into the pseudobulb for resources, it will lose its leaves and begin to shrivel up. Popular orchids that generate pseudobulbs include cattleya, cymbidium, oncidium, dendrobium, and epidendrum.

You can cut pseudobulbs from a mother plant with a section of stem attached, dip in rooting hormone, and pot up—similar to how other offsets are removed and potted. It is best to take pseudobulb cuttings in spring, when plants are most actively growing.

Bulbils, also called aerial bulbs, are another type of clone plantlet that develops on certain types

∧ Aerial bulbs, or bulbils, developing on a garlic flower stem.

of plants. They look like tiny bulbs and grow at the base of a stem, in the leaf axil of a plant, and atop flowering stalks—always above ground. These tiny structures can develop into an entirely new clone of the mother plant. It's like planting a seed that's a cutting! Bulbils are very easy to propagate once they've matured.

Bulbils are common in the onion and lily plant families. Garlic plants produce bulbils on flowering stalks,

Zygopetalum spp. produce pseudobulbs and their flowers are highly fragrant.

as do Egyptian walking onions. A few species of lilies produce bulbils along their stems, usually dark purple in color. Some agave species (which are in the lily family) also produce flower stalks that can be covered with many bulbils. To propagate a bulbil, simply bury it in rooting media just like you would a seed or bulb.

As long as a bulbil stays attached to the mother plant, it will continue to develop and grow larger. Eventually, most bulbils weigh down the stem or flower stalk and come in contact with the soil where they will root, or they detach and fall from the plant, rooting where they hit the soil. At this stage of growth, bulbils are commonly referred to as offsets.

FOLIAR EMBRYOS

Leaves of some plants, such as those in the succulent genera *Bryophyllum* (commonly known as mother of thousands) and *Kalanchoe*, will develop small plantlets attached along leaf margins (the edge of the leaf blade). Hen and chick ferns also produce tiny plantlets along their leaves. These small plantlets form from foliar embryos that are located at the edge of the leaf. Sometimes foliar embryos are also called bulbils.

As the leaf grows, so does the foliar embryo, until it develops one or two leaves with a stem and a tiny root. Then it stops growing and waits. These tiny plantlets won't develop any further until they come in contact with soil or water, either from the main leaf falling from the plant to the soil or the small plantlet breaking off and dropping.

Once in contact with some form of soil or water, the tiny plantlet begins to grow new root tissue and will then grow into a new plant. You can either detach an entire mother leaf and place it on some potting mix,

∧ This mother of thousands (*Bryophyllum daigremontianum*) plant is loaded with baby plantlets along the leaf margins, waiting to be brought to life.

or gently remove the tiny plantlets individually and set them on some moist growing media or potting mix. They will quickly root and begin growing.

Other plants, such as the piggyback plant (*Tolmiea menziesii*), generate new plantlets on top of the leaf surface. These leaves with attached plantlets can be cut from the new plant and placed in rooting media with the petiole buried.

Grape hyacinth (*Muscari* spp.) forms small true bulbs that multiply in clusters, creating dense clumps in the garden.

UNDERGROUND PROPAGULES

Now that we've covered the plant parts that typically grow above ground, it's time to dig a little deeper. There are several underground plant structures (propagules)—such as bulbs, rhizomes, and tubers—that you can use to multiply many plants.

BULBS

You should know about two types of true bulbs: tunicate bulbs, such as tulips, muscari, hyacinth, garlic, and onions, and non-tunicate (scaly) bulbs, such as lilies. You can vegetatively propagate either type. Bulbs have a central growing point (stem shoot), surrounded by modified leaf tissue that stores food and water. The shoot tip and the surrounding fleshy leaf tissue are attached to a stem that attaches to the base of the bulb.

A simple way to propagate bulbs is to let them naturally develop bulblets—a type of offset. Bulblets are small bulb-like structures that can develop at the base of tunicate bulbs, and on underground stem tissue of scaly bulbs. These tiny bulbs typically have root initials developing on them and can be gently cut or pulled away from the main bulb's structure. You can plant bulblets by submerging just the base into potting mix. They will continue to develop into full-sized bulbs, which will generate new plants. It will take several years for bulblets to mature to a flowering size.

If you want to encourage tunicate bulbs to produce more bulblets faster, you can damage the base of the bulb using several different techniques: coring out the center of the bulb, scooping out the base of the bulb, cross-cutting the base of the bulb, or cutting it into sections. Then dust the exposed areas with garden sulfur to prevent fungal growth. Place some slightly moist peat or coir in a plastic bag and insert your bulbs. Blow air into the bag to inflate it, then close the bag with a twist tie. Place in a dark cool-temperature area. In a few months you'll find tiny bulblets developing that you can remove and pot up.

To propagate scaly bulbs, simply remove the outer scales from the bulb, and submerge the bottom half into a container with potting mix or other rooting media. These scaly bulb leaves will produce bulblets.

Spring-blooming hyacinth is a type of true bulb. >

To force true bulbs to grow new bulblets, use one of these cutting/coring techniques. >

< These small bulblets grew from an older hyacinth bulb I left in a pot for 2 years. They can be divided and replanted, or you can let them mature where they emerge in the garden.

These are corms of
the saffron crocus
(*Crocus sativus*).
You can see baby
cormels growing
on top of the main
corm. >

CORMS

While often confused with bulbs, a corm is actually a swollen base of a flowering stem, wrapped in several thin leaves. Dense stem tissue forms inside these leaves. Plants that grow from corms, such as crocus and gladiolus, produce a new corm at the base of each new shoot, every year.

To propagate directly from corms, you'll need to dig up some in the fall when there are still green leaves left on the plant. Lay the clump on newspaper to dry completely. Then clean all the soil off the corms. You'll find tiny baby corms, called cormels, growing attached to the main corm—much like bulblets on a bulb. You can remove the cormels and plant them about 1 inch deep in growing media.

RHIZOMES

Rhizome cuttings are similar to root cuttings, except that a rhizome is actually a type of stem tissue (not root tissue) that grows underground. Ginger and canna lily plants, for example, grow underground rhizomes. Some plants, such as iris, have rhizomes that grow partially submerged. Other plants, such as rabbit's foot fern (*Davallia* spp.), have rhizomes that grow on top of the soil. Sympodial orchids—such as dendrobium, cattleya, and dracula orchids—produce offsets (keikis) from surface rhizomes.

This dracula orchid has filled out its original container and is producing new offsets atop surface rhizomes. I can snip a piece of rhizome and root it or remove the baby keikis and root them. ∧ >

⌐∧ Snake plant (*Sansevieria* spp.) grows thick rhizomes under the soil surface, which produce offsets that eventually push through the soil. You can cut these pups away from the mother plant with a piece of rhizome and some roots attached. Cut off the large leaves and then pot up. You can also cut sections of the rhizomes and pot them up before they've made new shoots. The latter would be considered a rhizome cutting.

A rhizome has nodes that produce both roots and stem shoots. When you cut a rhizome into separate pieces, each piece can grow roots and shoots and develop into a whole new plant. Plants that produce rhizomes and that can be propagated from rhizome cuttings include bamboo, ginger, begonias, canna lilies, potatoes, irises, asparagus, geraniums, hops, snake plant, some aquatic plants, and Venus flytrap, to name a few.

Many plants that produce offsets (pups) beneath the soil are doing so by underground stem tissue or from rhizomes. Eventually, when the rhizome growth hits the soil surface, it will develop a new shoot that will emerge as a pup. Before this happens, you can dig up rhizomes and cut them up to make multiple new plants.

How to Take a Rhizome Cutting

1 Disinfect your knife using alcohol or a 10 percent bleach solution.

2 For underground rhizomes, pull soil away from the base of the plant, exposing a rhizome.

3 Chose a section of rhizome that has started to develop some roots and has at least one bud node. Cut a 1- to 3-inch rhizome section. Allow sections to cure for several days so exposed cuts are dry. Dip into powder rooting hormone.

4 Place the rhizome cutting horizontally on top of or into moist growing media in trays or 4-inch pots and cover lightly with more growing media. Keep the growing media moist but not wet. If you are using a plant that grows rhizomes on top of the soil, such as certain ferns, you'll set the rhizome cutting on top of the growing media and press down so it is in firm contact with the soil. You may need to use toothpicks or wire to keep it flush with the soil.

5 You can use a humidity dome or bag if you have trouble keeping the growing media moist to the touch, but underground rhizome cuttings often won't require a cover. Rhizomes that sit on top of the soil should be covered for a short time.

6 Rhizome cuttings won't require any exposure to light until they begin to sprout new shoots.

7 Once new shoots emerge, follow steps 7 to 9 in "How to Take a Stem-Tip Cutting" (page 144).

∧ Offsets can develop on iris rhizomes, which you can remove and pot up just like other offsets or pups.

An iris rhizome dug from the garden. You can cut the rhizome into several pieces, each including some bud and root tissue. By cutting the rhizome, the new pieces will be stimulated to grow new roots and shoots. Let the rhizome cuttings cure for several days, then replant them in the ground or into pots. >

∧ This sweet potato is beginning to sprout new shoots from buds called eyes. The tuber can be cut into several sections, each with at least one eye. Allow sections to cure for several days before planting.

TUBERS

Tubers are a type of underground swollen stem or root tissue used for resource storage. Potatoes are tubers, as are sweet potatoes, cassava, cyclamen, and dahlias. Tubers develop "eyes," which are shoot buds that will grow and develop new roots.

You can propagate tubers by cutting them into sections that each have at least one bud or eye. You'll need to cure these tuber sections, so the cut areas can dry and heal for several days before you place them back in the ground or in growing media. It's also good practice to dust your tuber cutting sections with garden sulfur before you replant them, to help cut down on fungal diseases.

∧ This piece of root I cut from a coneflower (*Echinacea* spp.) has sprouted a new shoot that can be potted up.

ROOT CUTTINGS AND SUCKERS

Some plants can develop new adventitious roots and shoots from a piece of their main root tissue. We call these root cuttings. Plants with thick, fleshy roots—including garden perennials such as *Anemone* spp., *Echinacea* spp., *Echinops* spp., *Eryngium* spp., *Gaillardia* spp., *Geranium* spp., *Farfugium* spp., *Mentha* spp., *Papaver* spp., *Phlox* spp., and *Salvia* spp.—are best suited to root cutting propagation. You can also propagate some shrubs (such as hydrangea) and a number of tropical trees by root cuttings.

A root cutting is just what it sounds like: you cut away a small section of root from the main root system, much like a cane cutting, and then pot it up into a small container of growing media. After several months the root cutting will develop new adventitious roots and new shoots. It's best to take root cuttings in winter or early spring when plants are still dormant.

How to Take a Root Cutting

1 Disinfect your knife using alcohol or a 10 percent bleach solution.

2 Pull soil away from the base of the plant, exposing part of the roots. Choose a section with roots each the thickness of a heavy gauge wire. Some plants will have thicker roots, say, the width of your pinky finger.

3 Cut a length of root, then cut it into sections 1 to 2 inches long. Dip the sections into the rooting hormone.

4 Place the root cuttings horizontally (or vertically with the cut side facing up and the natural base of the root section pointing down) on top of or into a moist growing media in trays or 4-inch pots and cover lightly with more growing media—about ½ inch deep. Keep the growing media moist but not wet.

5 You can use a humidity dome or bag if you have trouble keeping the growing media moist to the touch, but root cuttings often won't require a cover if soil is moist.

6 Your root cuttings won't need any exposure to light until you see bud shoots emerging from the soil.

7 Once new shoots emerge, follow steps 7 to 9 in "How to Take a Stem-Tip Cutting" (page 144).

∧ This blackberry sucker has emerged in the middle of my lawn, several feet away from the mother plant. I've pushed away some soil, exposing the roots emerging from the sucker. You can cut below these roots to remove the sucker and replant it.

Some plants develop what we call suckers from their root tissue. These are sprouts that emerge from roots under the soil. The new plantlets can be dug and potted up to multiply your plants. Suckers are often seen as a negative thing in the landscape, because they can pull energy away from your main plant. They can also be a nuisance as they will pop up in your lawn and landscape in undesirable areas. However, if you want to make more of the plant that is making suckers, then you're in luck. Blackberry and raspberry plants commonly sprout suckers in the landscape.

To propagate a sucker, simply follow the same process for taking an offset cutting. Carefully dig around in the soil around the sucker and cut it away from the main root system, making sure to include a portion of its root mat. Immediately plant the sucker elsewhere in your garden, or into small containers.

SIMPLE DIVISION

While we have already talked about dividing plants in different ways by removing assorted pieces and parts, we haven't yet covered simple division. Many plants grow in clumps that can be divided in half or several sections simply by cutting through the entire root ball. When you pull the entire clump apart, you have two or more fully functional plants ready to be transplanted to another place in the garden or potted up into containers.

If your plants grow by clumping, sending up offsets around the main central growing crown, or with clumps of rhizomes or bulbs, you can divide the entire clump by simply loosening the connections between the root system or other structures. Pull them apart to create separate root balls or free up whole bulbs, tubers, corms, or rhizomes for replanting. Division is the fastest way to make more of a mature plant you already have, if its growth habit allows for simple division.

Many herbaceous garden perennials and herbs, such as ornamental grasses, salvia, succulents, and mint, to name a few, are propagated using simple clump division. Bulbs and rhizomes that multiply in clumps, such as irises and daffodils, can also be divided by splitting clumps. You can also easily propagate certain orchids, ferns, and many tropical plants by simple division.

For plants with tight clumps and root systems, such as grasses, sedges, and edibles such as asparagus, you'll need to use a sharp knife or shovel to cut the clumps apart. For other looser clumping plants, such as succulents and other fleshy perennials such as pincushion flowers, you can often simply pull or tease apart the clumps by hand.

When a plant gets too big for its britches, outgrowing its space in the garden or in a container, division is a common remedy. When a clumping perennial has gotten too large for the allotted area you can divide it and replant a smaller clump—or move extra clumps to other parts of the landscape. Many bulbs that grow in clumps, such as daffodils, may require division every few years to counter overcrowding, which would result in fewer flowers.

There are a few different ways you can approach dividing your plant. If it is a perennial in the garden, you can use a sharp shovel or trowel and cut directly into the clump, then remove a section you want to

∧ This haworthia clump can be divided into separate whole plants. Each can be potted up into its own pot.

transplant, leaving part of the original clump in the ground. You also can dig up the entire clump, cut it into several sections, and then replant in different locations or into containers. If you are dividing a container plant, you'll remove it entirely from the container, cut or pull sections of the clump apart, then repot the new divisions into separate containers.

With these basic plant propagation categories and methods, you're ready to get started making more of many types of plants. As you get to know your plants better, you'll discover which methods of propagation work best for them, and you.

Many perennials, such as ornamental grasses, can be dug, divided, and replanted.

∧ > This African violet plant has multiple crowns and has
become crowded in its pot. I can divide the clump into individual whole plants and repot them separately.

Many pests and diseases want to make a meal out of your plants. I found this meaty tomato hornworm munching on my tomato plants.

Common
PESTS
and
DISEASES

How to Care for Your Plants

Once your seedlings emerge and begin developing new leaves, or you begin potting them up into larger containers, different pests or diseases can often crop up. It's also easy to unwittingly introduce a pest or disease when you bring outdoor plants inside for the winter or buy new plants for your indoor garden. These hitchhikers can quickly find a home and reproduce on your young seedlings and cuttings. You can also accidentally perpetuate a disease or pest problem if you take cuttings from an infested or infected mother plant. If you're keeping houseplants indoors, several pesky pests can become a recurring nuisance.

Even experienced plant growers will have issues with insects and diseases, especially when growing indoors. As previously discussed (see page 113), seeds that are germinating can succumb to damping off disease. But as seedlings grow, new predators can move in. For example, my pepper plant seedlings and mature plants often end up with aphids when I grow them indoors—even if the plants are well cared for and healthy. Outdoors, on the other hand, you'll rarely have a problem with aphids on peppers. The indoor conditions are more favorable for an aphid explosion.

If you grow your houseplants in soil with some organic matter, you will probably always have some fungus gnats, off and on. If you grow certain kinds

∧ These scale insects have covered several branches on an indoor lemon tree, sucking the life out of the plant.

of plants that are particularly susceptible to certain pests, such as citrus (spider mites, scale, and mealy bugs), you're bound to see these pests crop up from time to time.

Don't get discouraged. Just watch out for early signs of problems so you can keep them at bay. If you lose some plants to a pest or disease issue, it's no biggie—that's why you have this book—you can make many more!

< Aphids are a common pest problem on pepper seedlings and large plants as they are growing indoors. You can typically use soapy water or an insecticidal soap to rid your indoor plants of aphids. However, once you move the plants to the outside garden or patio containers, the aphids usually disappear naturally thanks to the environmental conditions and insect predators.

< Left untreated, aphids and whiteflies did a lot of damage to my young citrus cuttings, causing them to lose many of their leaves and flower buds.

Common Pests on Plants

PEST	DESCRIPTION	TARGET PLANTS	SIGNS/SYMPTOMS	TREATMENT
Aphids	Small, oval shaped, green, black, or white. Aphids suck water and nutrients from leaves and stems. They also carry viruses they pass from plant to plant as they feed.	Annual flowers, beans, beets, bok choy, cannabis, chard, citrus, cucumber, fruit trees, herbs, lettuce, peppers, perennials, and roses.	Curled or wilted foliage, stunted growth. General decline. Affected plants become more susceptible to other pests and diseases.	Insecticidal soap, horticultural oils, spinosad. Wash off leaves and stems with spray of water, or manual removal (squish them!).
Fungus gnats	Tiny black gnats that fly around the potted plant. Small white larvae in the soil.	Any potted plants growing indoors in potting soil that contains organic matter.	Larvae feed on the plant roots in the soil. Gnats flying about the plants and room. General plant decline.	Boost air circulation with fans. Reduce overwatering. Sticky traps. 10% hydrogen peroxide soil drench, Bt granules.
Mealy bugs	Soft-bodied, wingless, fuzzy cottony masses, some with long tails. On leaves, stems, and bark. They jump!	Cannabis, citrus, many foliage houseplants, gardenia, and succulents.	Yellowing and curling of leaves. Sticky honeydew residue.	Wash off, manual removal (squish!). Insecticidal soap, horticultural oils, spinosad. Systemic insecticide. Predatory insects. Move plants outside.
Scale	Oval-shaped insects with either a hard or cottony shell, with different colors. Cluster on stems and base of leaves.	Avocado, basil, citrus, many foliage houseplants, and succulents.	Stunted growth. Weak appearance, shriveled and yellow leaves that often drop off. Sticky honeydew residue. Fungus can grow on honeydew.	Wash off, manual removal. Insecticidal soap, horticultural oils, spinosad. Predatory insects. Systemic insecticide. Difficult to treat because of hard outer shell. Move plants outside.
Spider mites	Fast-moving, tiny arachnids with red-brown or pale coloring. Underside of leaves and stems. Fine webbing on leaves.	Annual flowers, basil, berries, cannabis, citrus, English ivy, foliage plants, herbs, perennials, roses, shrubs, tomatoes, and tropicals.	Pale or yellow mottled cast to leaves, leaf curling and dropping. Tiny pinprick holes on leaves.	Insecticidal soap, horticultural oils, spinosad. Predatory insects. Miticide. Persistent and require multiple treatments. Move plants outside.
White-flies	Tiny white flies on leaves and stems. If you shake the plant, they will fly around the plant. Larvae and nymphs suck plant sap.	Cannabis, citrus, cucumber, many foliage houseplants, grapes, potatoes, squash, strawberries, tomatoes, and tropicals.	Mottled leaf appearance, yellowing and dropping leaves, overall reduced growth and vigor. Sticky honeydew followed by sooty mold.	Sticky traps, insecticidal soap, horticultural oils, spinosad. Persistent and require repeat treatments. Move plants outside.

Common Plant Diseases

DISEASE	DESCRIPTION	TARGET PLANTS	SIGNS/SYMPTOMS	TREATMENT
Bacterial stem rot (*Erwinia*)	A bacterial disease that enters plant tissue through wounds in stems and leaves.	African violets, annual flowers, carrots, eggplant, foliage plants, peppers, philodendron, potatoes, tomatoes, and more.	One or two branches wilt first, with water-soaked black lesions on stems. The rest of the plant may then wilt and die.	No effective chemical control. Use clean cuttings and tools; keep workspace clean. Remove infected plants immediately. Avoid plant injury.
Early blight (*Alternaria*)	A leaf spot fungus spread by splashing rain, irrigation, insects, and garden tools. Common in humid, warm conditions.	Eggplant, peppers, potatoes, and tomatoes.	Lower leaves develop small brown spots. Spots spread, turning the leaf yellow, which then curls and drops.	No overhead irrigation or misting. Water at the soil level. Remove infected leaves immediately and improve air circulation. Foliar fungicide.
Damping off and root rot (*Pythium, Rhizoctonia*)	Damping off: Fungal pathogens that rot seedlings right at or just below the soil line. Root rot: Many pathogens that attack root systems, turning them brown.	All seedlings and young transplants. Any plants growing in waterlogged soils can suffer from root rot. Can be an issue in hydroponic systems.	Seedlings fall over at the soil line before they can mature. Plants begin to droop, wilt, turn brown, and collapse. Root tissue is brown and can be slimy.	Don't overwater seedlings or keep humidity too high. Use a sterile soil mix and manage soil temperature. Improve soil drainage and aeration. Oxygenate hydrosystems. Add beneficial bacteria.
Gray mold (*Botrytis*)	A mold disease that grows on leaf surfaces, blocking light from the leaf surface and causing severe damage to foliage and flowers.	African violets, berries, cannabis, carrots, flowering annuals, foliage plants, bulbs, grapes, peas, roses, tomatoes, and many more.	White spots on leaves or stems that turn gray, then brown. The fungus can cover large areas or the entire plant with webbing.	Good soil drainage. Improve air circulation between plants. Remove infected leaves and blooms. Foliar fungicide.
Powdery mildew (*Podosphaera*)	A fungal disease with fuzzy white growth. Spreads quickly, covering foliage and blocking photosynthesis.	Annual flowers, beans, cannabis, citrus, cucumbers, foliage plants, herbs, roses, peas, peppers, squash, tomatoes, and many more.	Powdery white substance on foliage. Reduced overall growth and vigor, stunted yellowing leaves that drop.	No overhead irrigation or misting. Control humidity levels. Water at the soil level. Foliar fungicide.
Sooty mold (*Alternaria, Cladosporium*)	Several types of fungal species. A gray- to black-colored mold that grows on the honeydew residue produced by aphids, scale, and whiteflies.	Any plant susceptible to aphids, scale, and whiteflies. Annual flowers, citrus, herbs, and perennials.	Mold spreads across the leaf surface, blocking light and photosynthesis. Leaves yellow and drop. Overall reduced growth and vigor.	Wipe off plant leaves with soapy water, a plant wash product, or neem oil. Control aphids, scale, and whiteflies.

TECHNIQUES FOR ELIMINATING PESTS AND DISEASE

～～～

Prevention is always the best medicine. It's good propagation policy to scout proactively for signs of diseases and insects. Never take cuttings from diseased plants and always inspect new plants you bring into your house. Consistently mind your watering habits as overwatering can quickly lead to soil-borne fungal diseases or encourage pests such as fungus gnats. Under-watering can trigger infestations from pests such as spider mites.

Clean your pots, tools, and seed trays between uses with a 10 percent hydrogen peroxide or bleach solution, or diluted vinegar. Diluted vinegar is also great for removing built-up salts from pots. If you have pruned an infected plant, sterilize your pruners in a 10 percent bleach solution before you use them on healthy plants.

Certain pests become a bigger problem when the air in your home is stagnant. Good ventilation and air movement are easy ways to reduce certain insects and diseases. If you struggle with fungus gnats or powdery mildew, you can use small fans to keep air moving around your seedlings and cuttings or place susceptible plants near air vents in your home where air may move more vigorously.

If you have a plant or plants that have become heavily infested with insects or infected with a disease, you should consider removing them completely from your growing area. Insects and diseases will potentially spread to your healthy plants if not separated. Sometimes, simply setting an infested plant outdoors on your patio or balcony for a time can naturally resolve the infestation. There are many natural predators outdoors, as well as wind, that can help cure your plant.

If you find insect eggs or disease spots on just one or two plant leaves, you can remove these leaves and dispose of them immediately. This may help stop the spread so you don't have to sacrifice the entire plant.

I often use sticky cards around indoor plants, and cuttings and seedlings, to trap and kill flying pests such as fungus gnats and whiteflies.

∧ Citrus cuttings and mature plants are particularly susceptible to sooty mold, which can completely cover leaves of infected
plants. You can use some soapy water to gently scrub the sooty
mold off the leaves.

∧ Mealy bugs cluster together at the ends of stem tissue and will hop away if you disturb them. Mealy bugs commonly infest many tropical houseplants, as well as citrus plants. Getting rid of mealy bugs will require a bit more patience and repeated treatments. If you catch them, you can squish them.

TREATMENTS

If you need to treat a plant for a pest or disease, it's wise to be mindful about the products you use, especially if you are treating edible plants, such as lettuce or tomatoes. I recommend avoiding systemic insecticides or fungicides on edible plants. Systemic products are absorbed into the plant tissue and can persist for extensive time periods. Even if you aren't growing edibles, if you're propagating plants to keep in your home, you probably also want to minimize chemical exposure. There are many organic and low-impact, topical treatment options.

Sometimes, all you need is a strong spray of water to wash off and kill certain insects, such as aphids and mites. Rinse plants under a running faucet or use a spray bottle and wipe down the leaves and stems. This method can remove a lot of the insects.

Insecticidal soaps, neem oil, and other types of horticultural oils are generally all you need to prevent or knock out pests on young seedlings or cuttings indoors. Treatments with spinosad (a natural chemical found in bacteria) are also very effective for killing chewing pests.

Thuricide (*Bacillus thuringiensis* or Bt) is an excellent natural treatment for chewing caterpillars—such as tomato hornworms or cabbage loopers—when sprayed in liquid form. Although, worm pests won't typically be a big problem for you when starting seedlings or cuttings indoors. In its dried crumbled form, Bt can be sprinkled on top of the soil, or mixed into the soil, of indoor houseplant or seedling pots to help control fungus gnat larvae.

Sometimes, beneficial predatory insects will show up to do the job for you—even in the indoor garden. Ladybugs and ladybug larvae are aphid destroyers. When

∧ You can use natural insect sprays, such as those that contain spinosad, on both edible and ornamental plants to control a variety of insects.

you see them on your plants, that usually means a pest insect is already there as a food source. Don't kill them!

If your plants have a fungal disease problem, look to foliar sprays with an organic fungicide, copper fungicide, neem oil, or plant wash. These are effective against diseases such as powdery mildew. Potassium bicarbonate is also a good treatment for a variety of fungal diseases.

No matter what you put on (or in) your plants, be it a treatment or fertilizer, always read the label before you use it and follow the label application instructions.

Healthy plants are better equipped to fend off insect and disease problems. Good plant care, adequate lighting, the right temperature, and even watering practices are your best defense against plant pests and diseases.

These ladybug larvae hatched on some of my indoor citrus plants. They are munching on aphids and whiteflies! >

My young begonia cutting is growing
on nicely in a 6-inch ceramic pot.

GRADUATION
Time

**When and How to Transplant Your
Seedlings and Cuttings**

I germinated these assorted cactus and succulent seeds directly in tiny clay pots, where they will stay for quite some time. ⟩

After your seeds successfully germinate, or your cuttings take root, development time will vary based on the type of plant you're growing. Some plants, especially those endemic to tropical environments, will grow and develop quickly after taking root. These fast growers will need to be potted up fairly soon after rooting.

Plants that are endemic to extreme climates, such as cacti and succulents, where seeds must germinate very quickly to survive—or leaves or stem pieces must root quickly—may thereafter grow very slowly, as resources in their natural environment are limited. Slow growers may stay in small containers for a long time before they are ready to bump up to a larger one.

∧ These whole-leaf succulent cuttings have grown baby shoots, so I'm planting them in tiny clay pots. The mother leaf will die away, and the plantlet can grow in the small pot until its roots fill the container.

< A few months after germination, this *Echeveria* 'Laui' seedling is still very tiny. It will remain in a tiny container for several more months before a large enough root system develops and it can be moved to a bigger container.

WHEN TO BUMP UP YOUR PLANTS

~

It's all about good timing when it comes to transplanting, or bumping up, your plants. If you're not sure how to gauge when a young plant is ready for transplanting to a larger container, the root system will tell you. Once the roots of your seedlings or cuttings have grown to the edges of their seed plug or pot, and the seedling has sprouted three or more true leaves (or the cutting has begun to put out new foliage), it is time to bump them up to larger containers. If you're water rooting or growing cuttings in an aeroponic propagator, your cuttings are ready to be potted up when roots are 1 to 2 inches long and begin branching.

Two common mistakes that new plant propagators make include transplanting young seedlings or clones before they've developed a large enough root system—or waiting too long to transplant them. If you transplant your young seedling or cutting too early, it could fail, since the root system isn't big enough to handle the disruption or a bigger pot with more soil.

On the other hand, if you wait too long to transplant your seedlings and cuttings, they can become stunted from overgrowing in their seed plug or cell and may never recover. If you leave water-rooted cuttings in water too long, allowing your plants to develop a large root system, they can have trouble transitioning to soil. Growth can be stunted for a while as plants acclimatize to their new growing environment. Either way, we call this phenomenon transplant shock.

True leaves on seedlings look like a miniature version of mature leaves. Once they emerge, the process of photosynthesis begins; the seedling is no longer relying on reserves. That's my signal to check their root growth to see if they are ready to transplant.

When you use pellets, seed plugs, or soil blocks, it allows the seedling to grow in the small casing until the roots hit the edge of the netting, newspaper, or soil block. That's when you know your seedling or cutting is ready to transplant to a larger container.

These echeveria tip cuttings have rooted into their pots and are ready to be transplanted into larger pots

∧ The cutting on the right is just about ready to transplant to a container with growing media. I kept the cutting on the left in the propagator a little longer than necessary. It will probably end up growing okay, but it might initially experience a bit more transplant shock.

∧ A rooted citrus cutting potted up into a 4-inch plant pot.

Scale up your container sizes in phases—don't go too big too soon. Generally speaking, you should replant a 2-inch seedling plug or cutting into a 4-inch diameter pot—not a 1-gallon pot. Once the root system has filled out the 4-inch container, then you can graduate it to a larger pot, its final container, or outdoor garden destination. If you plant a very small seedling in a much bigger container or into the outdoor garden, it may drown in too much water held by the larger soil volume or it could dry out as water moves away from the root ball to the edges of the container or in drying soil outdoors.

< These tomato seedlings have started developing their first set of true leaves. Time to check their root growth.

< After gently removing two plugs from the growing tray, I can see that the tomato seedling on the left has rooted completely to the bottom of the cell and is ready for transplanting. The plug on the right belongs to a slower-growing dwarf variety; its root system hasn't quite filled the entire plug and some soil has fallen away when I removed it.

I filled a 4-inch pot with a loose potting mix, then used a dibber to make a plug-sized hole for the tomato seedling. I dropped it into the hole, then gently pressed soil around the root ball and covered any exposed root area with some additional potting mix. I sprinkled some coir on top to help regulate moisture. ∧ >

The arugula seedling on the left has no roots emerging from the seed plug casing. The seedling below has rooted out of the casing and is ready for planting into a 4-inch container.

∧ Fast crops such as lettuce are rooted and ready to bump up to a 4-inch pot within just a couple of weeks of germination. After the plant roots fill the new pot, you can plant it in the outdoor garden, or move it to a 1-gallon container for indoor growing.

While many cactus seeds can germinate quickly, seedlings can remain small for years. It will take quite a long time before this baby is ready to graduate to a larger pot. ❯

These basil seedlings have developed new true leaves and have rooted to the edges of their seed plugs. They are ready to be potted up into 4-inch containers. Once the transplants have roots that fill up the 4-inch pots, they can be planted in the garden, or transplanted to larger containers.

The seedlings of these super dwarf 'Red Robin' tomatoes are so small, even after developing true leaves and adequate roots, I often transplant them into 2-inch pots *before* potting them into a larger container.

HOW TO BUMP UP YOUR PLANTS

~~~

Most often, you'll do best to choose growing containers with drainage holes in the bottom, or porous fabric grow pots that allow water and air to permeate. Use water-catching trays or tubs to collect the excess water that drains from your pots. Many types of plants will drown if you plant them into a container without a hole. If you are a chronic overwaterer, stay away from containers without drainage holes. You can get away with planting into a solid container if you are growing either plants that love water and won't mind wet feet, or succulents or cactus plants that you will only water sparingly so that no excess water builds up in the pot.

Fill your new transplant pot with a quality potting soil or custom mix. If you're transplanting succulents, use a loose potting mix labeled for cacti and succulents. If you're transplanting veggies and flowers, you can use a general purpose, lightweight potting mix. Avoid using heavy potting mixes meant for shrubs or outdoor containers with small transplants. You can always mix in some coir to help loosen up a soil mix that might be too heavy for young plant starts. Excellent solid fertilizer

and soil amendments you can use in your planting mix include worm castings, kelp meal, greensand, lava sand, blood meal, bone meal, fish meal, crushed crab shells, and dry molasses. These amendments will help feed your plants naturally over time.

The next step is to gently transfer your seedling or cutting into the new pot at the same depth at which the seedling naturally emerges. Never pull on a seedling's stem without first completely loosening it from the sides of the container. Doing so can break the seedling away from its root system. If the seedling has a good root system, you may be able to squeeze both sides of the plug and the seedling will pop right out. If the seedling doesn't have much of a root system, it is probably too early to transplant.

If you're using a plug tray filled with soil or a soilless growing media, squeeze the bottom of the plug to loosen the roots away from the inside of the container, then push your finger up against the bottom of the plug (or partially through the drainage hole) to loosen and slide the seedling or cutting up and out of the tray without damaging it. You can also use a knife or spoon. Hold the seedling underneath the root ball to set it into its new container so too much soil doesn't fall away from the delicate roots.

If you're growing your seedlings or cuttings in encased seed plugs, Oasis, or rockwool, you won't have to worry as much about disturbing the root system. With seed plugs you simply lift them out of the tray and place them into the new container. Oasis plugs typically pop right out of their tray cells. And with rockwool sheets, you'll use a knife to cut out the seedling or cutting, then set it in the new container.

Backfill your new container with additional potting mix and lightly press it in and around your new seedling. Fill the soil line to the same level as where your seedling naturally emerged from its original container. Most plants will suffer from potential rotting if you bury them too deep, although some plants such as tomatoes can be planted deeper.

Now that your transplants are actively photosynthesizing and growing, they will take up water faster and can dry out quickly under grow lights. Check their water needs more frequently than when the tiny seedlings were safe under a humidity dome.

∧ Using a small trowel or dibber, make a hole in the growing media the same depth as the seedling plug. Lightly cover the surface of the plug.

< ∨ The herb cuttings I stuck in rockwool have rooted to the bottom of the tray. Each block can be cut or gently pulled away to be potted up.

∧ Use a small brush or dust blower to gently remove soil from atop delicate seedlings and transplants.

∧ Can you see the fuzzy part of the tomato stem that has a purple tint? You can bury that portion of tomato stem under the soil, as it will develop new adventitious bracing roots from that part of the stem. If you snipped this seedling off right at the soil line, you could also use it as a stem-tip cutting—the stem will grow new roots from the purple area.

GRADUATE TO THE GARDEN

~

Once the root system has filled in, reaching the edges and bottom of its new pot and is extensive enough to hold together most of the soil in the pot, your transplant is ready to be moved into a larger container—for continued growth indoors or outdoors—or planted into your outdoor garden (if temperatures are appropriate, of course). This can take anywhere from 5 to 10 weeks for seedlings or faster for root cuttings. The type of crop and the growth rate will dictate what size pot to choose. Again, don't go too big, too fast. Your transplant may require several size upgrades before it is large enough to be planted into its final container or garden location.

If you're transplanting your seedlings or cuttings to an outdoor garden, or into outdoor patio or balcony containers, realize there is no one-size-fits-all when it comes to the right time to plant different crops outdoors. Gardening is a *hyper-local* experience. Be sure to follow recommended outdoor planting dates

for your local area. Look up your first and last average frost dates and then keep an eye on weather as the seasons transition. Get to know a little more about your plants and their ideal growing temperature ranges before you make the move. Planting outside too early could mean losing your new plants to cold temperatures. Planting outside too late could mean it's too hot for your chosen crop.

For example, many gardening experts (in the Northern Hemisphere) will tell you that you should never plant tomatoes out in the garden until late April or May. You'll find the same information presented online or on the seed packet itself. That timing might work great if you live in a cold climate. However, if you live in a hot climate like I do in Texas, you're better off planting your tomatoes outdoors in late February or early March to beat the heat and get good fruit set. Tomatoes might be tropical plants, but they don't set fruit well when temperatures are very hot. That

∧ The watercress, rosemary, and thyme cuttings I took, are now growing in 4-inch pots. Once their roots fill out the pots, they're ready to be planted into the outside garden or larger containers.

means I must start my tomato seeds in January, so the transplants are big enough to move outside on time. If you live in a colder climate, you might wait until March to start your tomato seeds indoors, for transplanting outdoors in May. If you're gardening in the Southern Hemisphere, then of course flip your months and seasons. It's all about your local climate and temperature trends.

As you become more experienced with propagation and your indoor and outdoor growing conditions, you'll be better able to predict the best planting times for your plants and garden.

˅ Now that this tomato transplant has rooted out to all sides of the 4-inch pot, it is ready to be bumped up to a larger container or planted into the outdoor garden.

Once the tomato transplant is settled in its new permanent home—a large 5-gallon container (that I'll grow indoors in a grow tent with grow lights)—I can start my days-to-harvest clock.

∧ After this purple cabbage transplant filled out
its 4-inch pot, I planted it into my Texas vegetable garden beds in mid-October. I started these
seeds indoors at the beginning of September.
Depending on where you live, you might need
to start your seeds and plant your cabbage outdoors earlier in fall, or in early spring.

RESOURCES

So, are you now ready to make more of the plants you love? While we only scratched the surface of this big and technical topic, the basic plant propagation techniques and tips covered in this book will certainly get you started—and help keep you growing. Some plants will multiply easily for you, while others may present long-term challenges. As you experiment with each of the basic techniques we covered you'll find certain failures—and successes—it's all part of the fun of learning to be a good plant parent.

You can visit me online at lesliehalleck.com for my Plantgeek Chic blog and other gardening and horticulture information.

INSTAGRAM AND TWITTER @lesliehalleck

FACEBOOK facebook.com/Halleck Horticultural

FACEBOOK GROUPS "Gardening Under Lights" (for indoor gardening enthusiasts) and "Plant Parenting" (for plant propagation enthusiasts)—join both!

PINTEREST pinterest.com/lesliehalleck

YOUTUBE Leslie Halleck

LINKEDIN linkedin.com/in/lesliehalleck

< Assorted succulent whole-leaf cuttings set atop dry potting mix to cure.

SUPPLY SOURCES

There are many good sources, both in-person and online, for seeds, seed-starting supplies, grow lighting, tools, and plants. Be sure to check your local garden center or plant shop for supplies and educational classes. Here are a few online resources to get you started.

Propagation Supplies

Avoseed makes the cute avocado seed sprouting "boats", as well as other kitchen accessories. avoseedo.com

Design Sprout makes the handcrafted ceramic sprouting supports I used in some of the projects. designsprout.nl

Gardener's Supply Company offers a plethora of gardening supplies for both the indoor and outdoor garden, including small indoor growing systems, HO T5 LED grow lights, and more. gardeners.com

Garden Supply Guys, Inc. is a friendly retail indoor growing supply store with a comprehensive and easy to use website for online ordering seed starting supplies and grow lights. gardensupplyguys.com

Hawthorne Gardening Company, manufactures and distributes a wide variety of grow lamps and growing gear for the serious and subtle enthusiast, as well as educational content. hawthornegc.com

Lee Valley Tools offers a wide array of gadgets and tools for your indoor and outdoor garden. leevalley.com

Second Sun Garden Supply offers a wide variety of indoor gardening supplies and colorful growing pots. secondsungarden.com

Soltech Solutions, Inc. designs and sells attractive LED grow lamps for your indoor living spaces. soltechsolutionsllc.com

Seeds

Baker Creek Heirloom Seeds offers a wide variety of heirloom rare seeds. **rareseeds.com**

Jelitto Seeds offers a wide variety of unique and interesting seeds for the United States and Europe. **jelitto.com**

Johnny's Select Seeds offers a wide variety of seeds and seed starting supplies. **johnnyseeds.com**

Park Seed offers a wide variety of seeds and seed-starting supplies. **parkseed.com**

Seed Saver's Exchange offers a wide variety of heirloom seeds. **seedsavers.org**

Seeds of Change offers a wide variety of organic seeds. **seedsofchange.com**

Sustainable Seed Company offers organic and heirloom seeds. **sustainableseedco.com**

Territorial Seed Company offers a wide variety of hybrid and heirloom seeds. **territorialseed.com**

Check out **etsy.com** for a wide variety of unusual succulent and cactus seeds.

Plants

Annie's Annuals and Perennials has lots of unique annuals and perennials. **anniesannuals.com**

Brent and Becky's Bulbs offers an excellent selection of bulbs for many climates. **brentandbeckysbulbs.com**

Four Winds Growers is the place to go for dwarf and semi-dwarf citrus plants. **fourwindsgrowers.com**

Logee's offers tons of tropical houseplants and edibles. **logees.com**

Mountain Crest Gardens has a huge succulent selection—of both cuttings and plants. **mountaincrestgardens.com**

Seattle Orchid offers a huge orchid selection. **seattleorchid.com**

The Sill offers lots of easy-care houseplants for new plant parents. **thesill.com**

ADDITIONAL LEARNING

For many of you, this book covers all you'll ever need for your home gardening and plant-keeping projects. Some of you might be inspired to dig deeper and learn more about these specific techniques, and graduate beyond them to more advanced methods of plant propagation. Here are some websites and books that may be helpful on your propagation journey.

The American Orchid Society website offers up lots of information to get you started with orchids. aos.org

The Cactus and Succulent Society of America publishes in-depth information on these popular plants. cssainc.org

Small-Space Vegetable Gardens: Growing Great Edibles in Containers, Raised Beds, and Small Plots by Andrea Bellamy

Succulents Simplified: Growing, Designing, and Crafting with 100 Easy-Care Varieties by Debra Lee Baldwin

The Manual of Plant Grafting: Practical Techniques for Ornamentals, Vegetables, and Fruit by Peter T. MacDonald

The Manual of Seed Saving: Harvesting, Storing, and Sowing Techniques for Vegetables, Herbs, and Fruits by Andrea Heistinger

The Indestructible Houseplant: 200 Beautiful Plants That Everyone Can Grow by Tovah Martin

The Reference Manual of Woody Plant Propagation: From Seed to Tissue Culture by Michael A. Dirr and Charles W. Heuser, Jr.

Be sure to check out the entire Timber Press catalog of books for more resources on topics related to plant propagation. timberpress.com

BIBLIOGRAPHY

Dole, J., and H.F. Wilkins. 2004. *Floriculture: Principles and Species*. 2nd ed. London, UK: Pearson.

Hartmann, H.T., D.E. Kester, and F.T Davies. 1990. *Plant Propagation: Principles and Practices*. 5th ed. Englewood Cliffs, NJ: Prentice-Hall, Inc.

Kyte, L., J. Kleyn, H. Scoggins, and M. Bridgen. 2013. *Plants from Test Tubes: An Introduction to Micropropagation*. 4th ed. Portland, OR: Timber Press.

PHOTO AND ILLUSTRATION CREDITS

All photos are by the author, except for the following:

Steven Edholm, page 124 (top)
Dan Heims / Cannsult Ltd., page 33 (right)
iStock.com/LagunaticPhoto, page 185
iStock.com/stevelenzphoto, page 30
iStock.com/temmuzcan, page 183
Lisa Lefebvre of Designsprout.nl, pages 2 (top left and bottom left), 22 (top), 91, 92, 96, 175
Jesse Nelson, Garden Supply Guys, page 86 (top and bottom right)

Holly Scoggins, page 23
Hawthorne Gardening Company, pages 61, 81 (bottom row), 84, 159
Caroline de Testa, page 3

All illustrations by Cathleen Green.

ACKNOWLEDGMENTS

Writing a book, especially one with projects and photos involved, is a time-consuming process that requires leaning on others for support and resources. For helping me get this project completed, on top of helping me run my green industry consulting company, I must thank my accounts manager at Halleck Horticultural, LLC, and book assistant, Jill Mullaney—she's an expert juggler. And as always, Elizabeth Krause, who also works with me on various writing projects, for applying her laser focus to wrangle my writing. Many thanks to Tom Fischer and the entire hardworking team at Timber Press for the opportunity to create this book. And last, but certainly never least, everlasting thanks to my ever-patient and supportive husband, Sean Halleck, for yet again tolerating both my plant addictions and insane work schedule. You're all the best.

INDEX

© Stacey Jemison

LESLIE F. HALLECK is a certified professional horti-culturist (ASHS) who has spent her 25+ year career hybridizing horticulture science with home-gardening consumer needs. Halleck earned a B.S. in biology/bot-any from the University of North Texas and an M.S. in horticulture from Michigan State University. The focus of her graduate research was greenhouse plant production using environmental controls such as lighting, temperature, photoperiod, and vernalization. Halleck's professional experience is well-rounded, with time spent in field research, public gardens, landscaping, garden writing, garden center retail, and horticulture consulting. At the end of 2012 Halleck devoted herself full-time to running her company, Halleck Horticultural, LLC, a green industry horti-cultural consulting and marketing agency. Halleck's previous positions include director of horticulture research at the Dallas Arboretum and general manager for North Haven Gardens (IGC) in Dallas, TX. Halleck is now a regular feature on the professional speaking and industry publication circuit, but she continues to offer up commonsense gardening advice and hands-on learning to home gardeners via her Plantgeek Chic blog, public workshops, and consumer publications. During her career, Halleck has written hundreds of articles for local, regional, and national publications as well as taught countless gardening programs for the home flower gardener, edible enthusiast, and back-yard farmer. In 2018, she authored *Gardening Under Lights: The Complete Guide for Indoor Growers* (Timber Press)—an in-depth guide to grow lighting and grow-ing just about anything indoors.